AF410860

Circuits and Systems Advances in Near Threshold Computing

Circuits and Systems Advances in Near Threshold Computing

Editor

Sanghamitra Roy

MDPI • Basel • Beijing • Wuhan • Barcelona • Belgrade • Manchester • Tokyo • Cluj • Tianjin

Editor
Sanghamitra Roy
Utah State University
USA

Editorial Office
MDPI
St. Alban-Anlage 66
4052 Basel, Switzerland

This is a reprint of articles from the Special Issue published online in the open access journal *Journal of Low Power Electronics and Applications* (ISSN 2079-9268) (available at: https://www.mdpi.com/journal/jlpea/special_issues/NTC).

For citation purposes, cite each article independently as indicated on the article page online and as indicated below:

LastName, A.A.; LastName, B.B.; LastName, C.C. Article Title. *Journal Name* **Year**, *Volume Number*, Page Range.

ISBN 978-3-0365-0720-0 (Hbk)
ISBN 978-3-0365-0721-7 (PDF)

Contents

About the Editor

Sanghamitra Roy is an Associate Professor in the Department of Electrical and Computer Engineering at Utah State University. She received her Ph.D. degree in Electrical and Computer Engineering from the University of Wisconsin-Madison. She received her M.S. degree in Computer Engineering from Northwestern University in December 2003. Dr. Roy has authored over 80 peer-reviewed publications in top-tier journals and conferences, as well as three book chapters in VLSI Design Automation. She serves in the editorial boards of the IEEE Design and Test Magazine and the Journal of Low Power Electronics and Applications. She has also served in the technical program committees of DAC, DATE, ICCAD, and ESWEEK, among others. She has received Best Paper Award nominations at the IEEE Design Automation & Test in Europe (DATE), 2011; the IEEE/ACM International Conference on Computer Aided Design (ICCAD), 2005; the IEEE 23rd International Conference on VLSI Design (VLSI Design), 2010; the IEEE/ACM International Conference on Hardware/Software Codesign and System Synthesis, 2014; and the IEEE/ACM Design Automation and Test in Europe, March 2018. She won the Best Paper Award at the 30th IEEE International Conference on Computer Design (ICCD), 2012. She received the NSF CAREER Award in 2013. Her research interests are in VLSI circuit design and optimization, and exploring reliability-aware novel circuit styles and architectures. Dr. Roy was named in the "125 People of Impact"—the list of the most influential alumni to graduate from the University of Wisconsin-Madison—in recognition of her ongoing success in academia and research. She is the inventor of 12 issued US patents.

Preface to "Circuits and Systems Advances in Near Threshold Computing"

Modern society is witnessing a sea change in ubiquitous computing, in which people have embraced computing systems as an indispensable part of day-to-day existence. Computation, storage, and communication abilities of smartphones, for example, have undergone monumental changes over the past decade. However, global emphasis on creating and sustaining green environments is leading to a rapid and ongoing proliferation of edge computing systems and applications. As a broad spectrum of healthcare, home, and transport applications shift to the edge of the network, near-threshold computing (NTC) is emerging as one of the promising low-power computing platforms. An NTC device sets its supply voltage close to its threshold voltage, dramatically reducing the energy consumption. Despite showing substantial promise in terms of energy efficiency, NTC is yet to see widescale commercial adoption. This is because circuits and systems operating with NTC suffer from several problems, including increased sensitivity to process variation, reliability problems, performance degradation, and security vulnerabilities, to name a few. To realize its potential, we need designs, techniques, and solutions to overcome these challenges associated with NTC circuits and systems. The readers of this book will be able to familiarize themselves with recent advances in electronics systems, focusing on near-threshold computing.

Sanghamitra Roy
Editor

Review

Near-Threshold Voltage Design Techniques for Heterogenous Manycore System-on-Chips

Sriram Vangal *, Somnath Paul, Steven Hsu, Amit Agarwal, Ram Krishnamurthy, James Tschanz and Vivek De

Circuit Research, Intel Labs, Intel Corporation, Hillsboro, OR 97124, USA; somnath.paul@intel.com (S.P.); steven.k.hsu@intel.com (S.H.); amit1.agarwal@intel.com (A.A.); ram.krishnamurthy@intel.com (R.K.); james.w.tschanz@intel.com (J.T.); vivek.de@intel.com (V.D.)
* Correspondence: sriram.r.vangal@intel.com

Received: 14 April 2020; Accepted: 7 May 2020; Published: 14 May 2020

Abstract: Aggressive power supply scaling into the near-threshold voltage (NTV) region holds great potential for applications with strict energy budgets, since the energy efficiency peaks as the supply voltage approaches the threshold voltage (V_T) of the CMOS transistors. The improved silicon energy efficiency promises to fit more cores in a given power envelope. As a result, many-core Near-threshold computing (NTC) has emerged as an attractive paradigm. Realizing energy-efficient heterogenous system on chips (SoCs) necessitates key NTV-optimized ingredients, recipes and IP blocks; including CPUs, graphic vector engines, interconnect fabrics and mm-scale microcontroller (MCU) designs. We discuss application of NTV design techniques, necessary for reliable operation over a wide supply voltage range—from nominal down to the NTV regime, and for a variety of IPs. Evaluation results spanning Intel's 32-, 22- and 14-nm CMOS technologies across four test chips are presented, confirming substantial energy benefits that scale well with Moore's law.

Keywords: NTV; NTC; low-power; low-voltage memory and clocking circuits; minimum-energy design; power-performance; resilient adaptive computing

---

1. Introduction

Near-threshold computing promises dramatic improvements in energy efficiency. For many CMOS designs, the energy consumption reaches an absolute minimum in the NTV regime that is of the order of magnitude improvement over super-threshold operation [1–3]. However, frequency degradation due to aggressive voltage scaling may not be acceptable across all single-threaded or performance-constrained applications. The key challenge is to lock-in this excellent energy efficiency benefit at NTV, while addressing the impacts of (a) loss in silicon frequency, (b) increased performance variations and (c) higher functional failure rates in memory and logic circuits. Enabling digital designs to operate over a wide voltage range is key to achieving the best energy efficiency [2], while satisfying varying application performance demands. To tap the full latent potential of NTC, multi-layered co-optimization approaches that crosscut architecture, devices, design, circuits, tool flows and methodologies, and coupled with fine-grain power management techniques are mandatory to realize NTC circuits and systems in scaled CMOS process nodes.

The overarching goal of this work is to advance NTV computing, demonstrate its energy benefits, to quantify and overcome the barriers that have historically relegated ultralow-voltage operation to niche markets. We present four multi-voltage designs across three technology nodes, featuring many-core SoC building blocks. The IPs demonstrate wide dynamic power-performance range, including reliable NTV regime operation for maximum energy efficiency. Key innovations in NTV circuit design methods and CAD approaches for wide-dynamic range design, including optimizations to design methodology are highlighted.

A design summary for the four NTV silicon prototypes is presented in Table 1. The first design describes an Intel Architecture IA-32 processor fabricated in 32-nm SoC platform technology with 2nd generation high-k/metal gate transistors [4]. The chip is capable of reliable ultra-low voltage operation and energy efficient performance across the wide voltage range from 280 mV to 1.2 V. The CPU (Figure 1a) consists of a Pentium™ class IA-32 core [5] with superscalar in-order pipeline, dynamic branch prediction and 8 KB of separate L1 instruction and data caches. Core logic and memory blocks are powered by independent voltage domains to allow processor core and the memories (L1 caches + microcode ROM) to operate at their individual optimal power levels for best overall energy efficiency. This capability allows the IA core logic to aggressively voltage scale well beyond memory minimum operating voltage (V_{min}) limits. A multi-voltage, NTV-aware core design synthesis and performance verification (PV) methodology is employed with measured core operation down to 280 mV sub-threshold regime. A minimum-energy voltage optimum (V_{OPT}) is observed at 450 mV for the NTV-CPU, signifying 4.7× better energy efficiency over $V_{DD\text{-}max}$ (1.2 V) operation [6].

Single instruction, multiple data (SIMD) permutation operations are key for maximizing high-performance microprocessor vector data path utilization in multimedia, graphics, and signal processing workloads [7]. The second prototype is a wide dynamic range design (240 mV to 1.2 V), presenting an ultra-low-voltage reconfigurable 4-way to 32-way SIMD vector permutation engine [8] (Figure 1b) consisting of a 32-entry × 256b 3-read/1-write ported register file (RF) with a 256b byte-wise any-to-any permute crossbar for 2-D shuffle operations across a 32 × 32 matrix. The register file with three read ports and one write port is used for vertical shuffle. The permute crossbar is used for horizontal shuffle. The SIMD engine is fabricated in a 22-nm SoC platform technology featuring 3-D tri-gate and high-k/metal gate devices [9]. Tri-gate or FinFET devices offer steeper subthreshold swing and improved short-channel effects and offer better variability and energy efficiency at NTV. The engine incorporates vector flip-flops, stacked min-delay buffers, shared gates to average min-sized transistor variations, and ultra-low-voltage split-output (ULVS) level shifters to improve logic V_{min} by 150 mV, while enabling a 9× peak energy efficiency of 585 GOPS/W measured at 260 mV supply voltage and a temperature of 50 °C.

Table 1. A summary of four NTV-optimized silicon designs from Intel Labs.

	ISSCC 2012 [6]	ISSCC 2012 [8]	VLSI 2013 [10]	VLSI 2016 [11]
NTV Design	32-b ×86 CPU	SIMD Engine	2D NoC fabric	mm-scale MCU
Intel CMOS Technology	32-nm high-K/metal-gate	22-nm Tri-gate	22-nm Tri-gate	14nm Tri-gate
Die Area (mm2)	2	0.048	Router: 0.051 2 × 2 Mesh: 0.93	0.79
VDD range and VOPT (V)	0.28–1.2 (VOPT = 0.45)	0.24–1.1 (VOPT = 0.26)	0.34–0.85 (VOPT = 0.4)	0.308–1.0 (VOPT = 0.37)
Frequency range (MHz)	3–915	10–2900	0.5–297	67–1000
Energy @ VOPT Benefit	170pJ/cyclea, 4.7×	1.9pJ/cycle, 9×	36pJ/cycleb, 3.3×	17.18pJ/cyclec, 4.8×
Total on-chip memory	8KB I\$ + 8KB D\$	1KB Reg. file (RF) memory	None	8KB I\$ + 8KB DTCM + 64KB SMEM + 16KB BootROM

[a] Pentium BIST workload, [b] For 2 × 2 NoC, [c] MCU *always-active*, running AES encryption.

Packet-switched routers are communication systems of choice for modern many-core SoCs [12,13]. The third NTV design describes a 2 × 2 2-D mesh network-on-chip (NoC) fabric (Figure 1c) which incorporates a 6-port, 2-lane packet-switched input-buffered wormhole router as a key building block [10]. The resilient NTV-NoC and router incorporates end-to-end forward error correction code (ECC) and within router recovery from transient timing failures using error-detection sequential (EDS) circuits and a novel architectural flow control units (FLIT) replay scheme. The router operates across a wide frequency (voltage) range from 1 GHz (0.85 V) to 67 MHz (340 mV), dissipating 28.5 mW to 675 µW and achieves 3.3× improvement in energy-efficiency at 400 mV V_{OPT}. The NTV-NoC is fabricated in the same 22-nm SoC technology [9] and achieves 63% higher bandwidths at V_{OPT} over a non-resilient router, when operating beyond the point-of-first failure (PoFF).

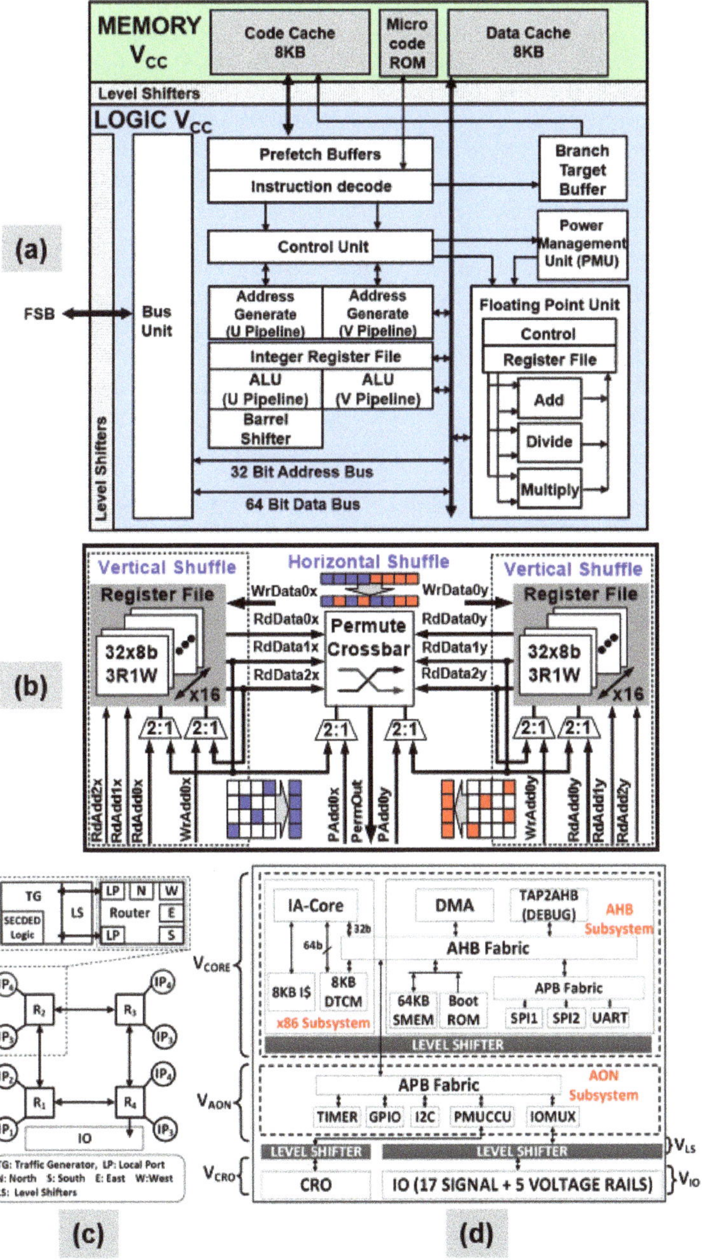

Figure 1. Block diagrams for four NTV prototypes: (**a**) Pentium™ class IA-32 CPU; (**b**) Reconfigurable 256-bit wide SIMD vector permutation engine with 2-D Shuffle; (**c**) Resilient four node (2 × 2) 2-D Mesh NoC fabric with routers (R_1–R_4); (**d**) mm-scale MCU with NTV Quark IA-32 core for µW WSNs.

The final NTV prototype showcases a wireless sensor node (WSN) platform that integrates a mm-scale, 0.79 mm^2 NTV IA-32 Quark™ microcontroller (Figure 1d) (MCU) [14,15], built using a 14-nm 2nd generation tri-gate CMOS process. The WSN platform includes a solar cell, energy harvester, flash memory, sensors and a Bluetooth Low Energy (BLE) radio, to enable always-on always-sensing (AOAS) and advanced edge computing capabilities in Internet-of-Things (IoT) systems [11]. The MCU features four independent voltage-frequency islands (VFI), a low-leakage SRAM array, an on-die oscillator clock source capable of operating at sub-threshold voltage, power-gating and multiple active/sleep states, managed by an integrated power management unit (PMU). The MCU operates across a wide frequency (voltage) range of 297 MHz (1 V) to 0.5 MHz (308 mV) and achieves 4.8× improvement in energy efficiency at an optimum supply voltage (V_{OPT}) of 370 mV, operating at 3.5 MHz. The WSN, powered by a solar cell, demonstrates sustained MHz AOAS operation, consuming only 360 µW.

This paper is organized as follows: Section 2 describes various NTV design techniques for SRAM and logic circuits. Architecture driven adaptive mechanisms to address higher functional failure rates and variation-tolerant resiliency at NTV for SoC fabrics are described in Section 3. Section 4 presents the tools, flows and recipes for wide-dynamic range design. In addition, solutions for multi-voltage global clock generation and distribution are introduced. Key experimental results from measuring all four prototypes are presented, analyzed, and discussed in Section 5. Finally, Section 6 concludes the paper and suggests future work.

2. NTV Circuit Design Methodology

The most common limit to voltage scaling is failure of SRAM and logic circuits. SRAM cells fail at low voltage because device mismatches degrade stability of the bit-cell for read, write or data retention. SRAM cells typically use the smallest transistors. Also, they are the most abundant among all circuit types on a die. Therefore, the V_{min} of the SRAM cell array limits V_{min} of the entire chip. Logic circuits, clocking, and sequentials fail at low voltage because of noise and process variations. Alpha and cosmic ray-induced soft errors cause transient failure of memory, sequentials, and logic at NTV. Frequency starts degrading exponentially as the supply voltage approaches V_T. This sets a limit on V_{min}. This limit can be alleviated to some extent by tri-gate transistors. Since they have a steeper sub-threshold swing, they can provide a lower V_T for the same leakage current target. Aging degradations cause failure of SRAM cells at low voltages since different transistors in the cell undergo different amounts of V_T shift under voltage–temperature stress and thus worsen device mismatches in the bit-cells. All these effects degrade and limit V_{min}. The following sections describe low-voltage design techniques used for SRAM memory, combinational cells, sequentials and voltage level shifters circuits.

2.1. SRAM Memory and Register File (RF) Optimizations

An 8-T SRAM cell (Figure 2a) is commonly used in single-V_{DD} microprocessor cores, particularly in performance critical low-level caches and multi-ported register-file arrays. The 8-T cell offers fast simultaneous read and write, dual-port capability, and generally lower V_{min} than the 6-T cell. With independent read and write ports in the 8-T cell, significantly improved read noise margins can be realized over the traditional 6-T SRAM cell, at an additional area expense. The noise margin improvement is due to the elimination of the read-disturb condition of the internal memory node by the introduction of a separate read port in the SRAM cell. As a result, variability tolerance is greatly enhanced, making it a desirable design choice for ULP SRAM memory operating at lower supply voltages down to NTV and energy-optimum points.

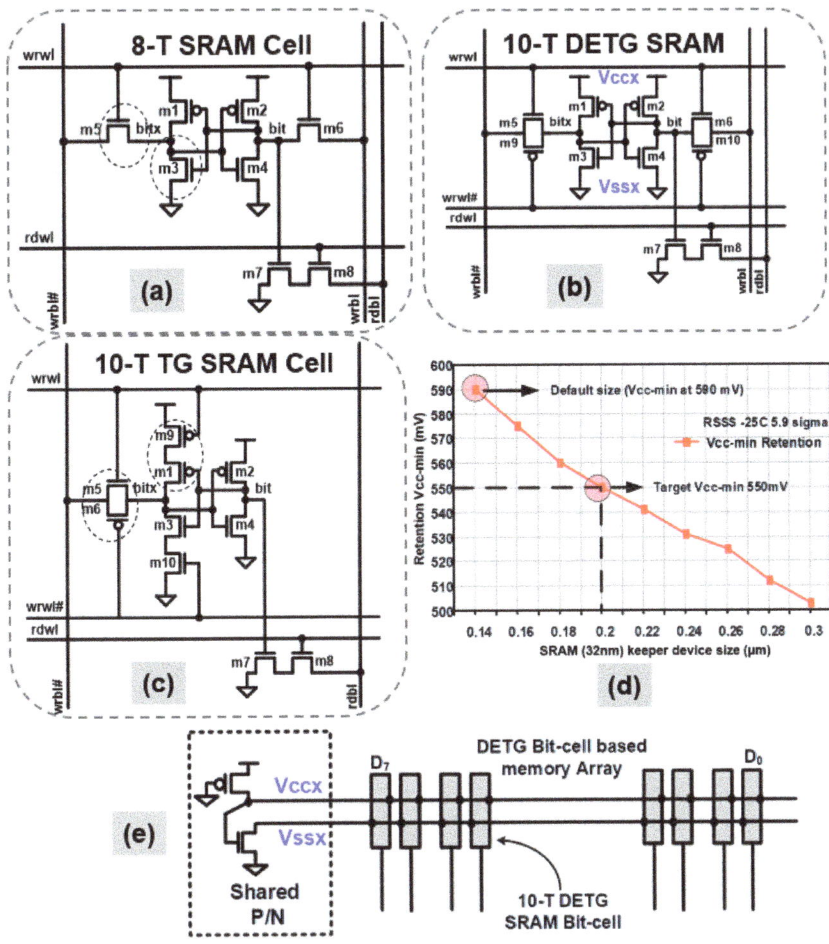

Figure 2. The prototypes use variability-tolerant SRAM bit-cells: (**a**) 8-T SRAM bit-cell used in the NTV-MCU; (**b**) The SIMD engine uses a 10-T dual-ended transmission gate (DETG) SRAM topology; (**c**) An alternate 10-T transmission gate (TG) SRAM bit-cell used in the NTV-CPU; (**d**) Simulated retention voltage simulations for the 10-T TG SRAM in 32-nm, as a function of keeper device size (m9, m10) in the presence of random variations (5.9σ, slow skew, $-25\,^{\circ}$C); (**e**) The shared PMOS/NMOS on the virtual supplies improve memory write V_{min} by 125 mV in the 22-nm DETG based memory array.

The 8-T bit-cell is still prone to write failures due to write contention between strong PMOS pull-up and a weak NMOS transfer device across PVT variation. This contention becomes worse as V_{DD} is lowered, limiting V_{min}. A variation-tolerant dual-ended transmission gate (DETG) cell is implemented on the 22-nm NTV-SIMD register file array by replacing the NMOS transfer devices with full transmission gates (Figure 2b). This design enables a strong "1" and "0" write on both sides of the cross-coupled inverter pair. The DETG cell always has two NMOS or two PMOS devices to write a "1" or "0", on nodes bit and bitx. This inherent redundancy averages the random variation effect across the transistors, improving both contention and write-completion. Moreover, the cell is symmetric with respect to PMOS and NMOS skew which reduces the effect of systematic variation. DETG cell simulations show 24% improvement in write delay, allowing a 150 mV reduction in write V_{min}. However, the DETG cell is contention limited at its write V_{min}, which can be reduced by the shared

P/N circuits. An always "ON" PMOS and NMOS is shared across the virtual supplies of eight DETG cells (Figure 2e). The shared P/N circuit limits the strength of the cross-coupled inverters across variations reducing write contention by 22%. This circuit optimization results in an additional 125 mV write reduction compared to DETG, enabling an overall 275 mV write V_{min} reduction when compared to the 8-T SRAM cell.

Caches in the 32-nm NTV-CPU use a modified, single-ended and fully interruptible 10-T transmission gate (TG) SRAM bit-cell (Figure 2c), which allows for contention-free write operations. This topology enables a 250 mV improvement in write V_{min} over an 8-T bit-cell. With this improvement, bit-cell retention now becomes a key V_{DD} limiter. The simulated retention voltage data for the 10-T TG SRAM, as a function of keeper device size (m9, m10) and in the presence of random variations (5.9σ, slow skew, $-25\,°C$) is shown in Figure 2d. Clearly, larger keeper devices lower the retention voltage. The keeper device is increased from 140-nm to 200-nm to realize a 550 mV retention V_{min} target. For reliable read operation, bit-lines incorporate a scan-controlled, programmable stacked keeper, which can be configured to three or four PMOS device stacks to reduce read contention and improve read V_{min}, across wide operating voltage/frequency range.

To achieve low standby power in the WSN, all on-die memories and caches on the 14-nm NTV-MCU use a custom 8-T (Figure 2a), $0.155\text{-}\mu m^2$ bit-cell, built using 84-nm gate pitch ultra-low power (ULP) transistors [14]. The 8-T bit-cell provides a well-balanced trade-off in V_{min} and area over the 6-T and 10-T SRAM cells. The ULP transistor optimized memory arrays are designed to provide low standby leakage. However, as summarized in Table 2, a 5× performance slowdown is estimated over standard performance (SP) transistor 8T memory at 500 mV, but is still fast enough for edge compute applications. Context-aware power-gating of each 2 KB array is supported for further leakage reduction with no state retention. The ULP array also enables 26× lower leakage (at 500 mV supply) and has a 55% area cost over an SP-based 8T memory array, drawn on a 70 nm gate pitch. The ULP memory leakage scales from 114 pA at 1 V voltage down to 8.28 pA per bit at the retention limit of 308 mV, as measured at room temperature ($25\,°C$).

Table 2. Comparison between 8T SRAMS build with ULP and SP bit-cells.

8T SRAM Device Type.	Gate Pitch	Normalized Frequency (0.5V)	Normalized Leakage (0.5V, 25C)	14 nm Bit-Cell Area (μm^2)
Standard performance (SP)	70nm	5×	26×	$0.100\ \mu m^2$
Ultralow power (ULP, NTV MCU Memory)	84nm	1×	1×	$0.155\ \mu m^2$ (1.55×)

Process, voltage and temperature (PVT) and aging adaptive on-die boosting of read word-line (RWL) and write word-line (WWL) as a common circuit assist technique for further lowering SRAM V_{min} is described in [16,17]. Boosting RWL enables larger read "ON" current without forcing a larger PMOS keeper. Boosting WWL helps write V_{min} for two reasons—it improves contention without upsizing NMOS pass device size (or lowering its V_{TH}), and improving write completion by writing a "1" from the other side. At iso-array area, on-die WL boosting achieves twice as much V_{min} reduction over bit- cell upsizing [16]. However, word-line boosting requires an integrated charge-pump, or another method for generating a boosted voltage on die.

2.2. Combinational Cells Design Criteria

Circuits are optimized for robust and reliable ultra-low voltage operation. A variation-aware pruning is performed on the standard cell library to eliminate the circuits which exhibit DC failures or extreme delay degradation at NTV due to reduced transistor on/off current ratios and increased sensitivity to process variations. Simulated 32-nm normalized gate delays (y-axis), as a function of V_{DD} for logic devices in the presence of random variations (6σ) is presented in Figure 3. Complex logic gates with four or more stacked devices and wide transmission-gate multiplexers with four or more inputs are pruned from the library because they exhibit more than 108% and 127% delay degradation compared to three stack or three-wide multiplexers respectively (Figure 3a,b). Critical timing paths are designed using low V_T devices because high V_T devices indicate 76% higher delay penalty at 300 mV supply, in the presence of variation (Figure 3c). All minimum-sized gates with transistor widths less than 2× of the process-allowed minimum (Z_{MIN}) are filtered from the library due to 130% higher variation impact (Figure 3d), and the use of single fin-width devices is limited in 22-nm and 14-nm logic design.

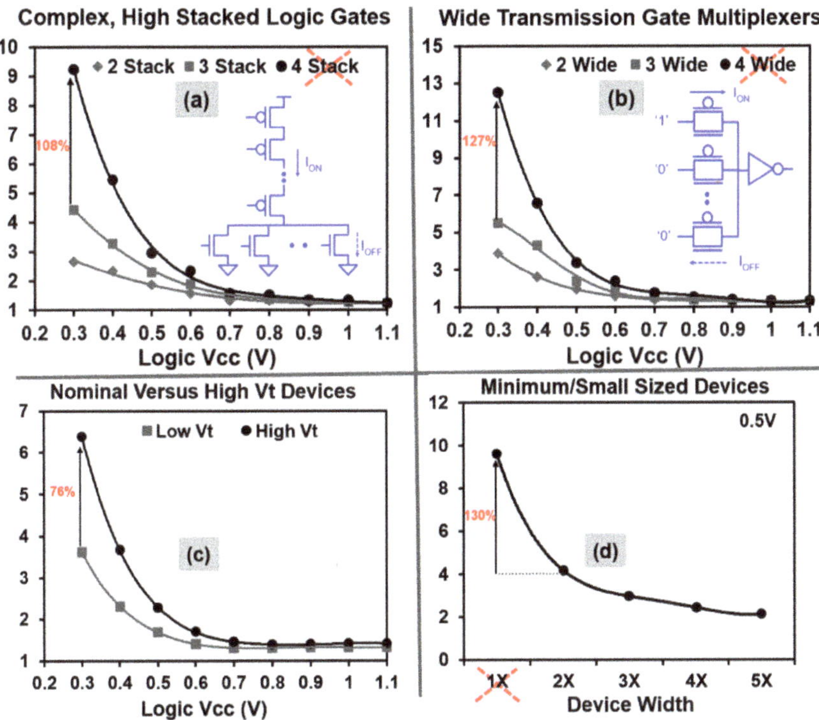

Figure 3. Simulated 32-nm normalized gate delays (y-axis) vs. supply voltage for logic devices in the presence of random variations (6σ). To limit excessive gate delays at NTV, the data indicates that: (**a**) Transistor stack sizes need to be limited to three, including; (**b**) Wide pass-gate multiplexers; (**c**) High V_T devices have 76% higher delay penalty over nominal V_T flavors due to variations; and (**d**) Minimum width (1×, Z_{MIN}) devices show 130% higher delay at 500 mV, requiring restricted use.

2.3. Sequential Circuit Optimizations

At lower supply voltages, degradation in transistor I_{on}/I_{off} ratio, random and systematic process variations, affect the stability of storage nodes in flip-flops. Conventional transmission gate based master-slave flip-flop circuits typically have weak keepers for state nodes and larger transmission gates. During the state retention phase, the on-current of weak keeper contends with the off-current of the strong transmission gate affecting state node stability. Additionally, charge-sharing between the internal master and slave nodes (write-back glitch) can result in state bit-flip due to reduced noise margins at low V_{DD}. The NTV-CPU employs custom sequential circuits to ensure robust operation at lower voltages under process variations. A clocked CMOS-style flip-flop design (Figure 4) replaces master and slave transmission gates with clocked inverters, thereby eliminating the risk of data write-back through the pass gates. In addition, keepers are upsized to improve state node retention and are made fully interruptible to avoid contention during the write phase of the clock, thus improving V_{min}.

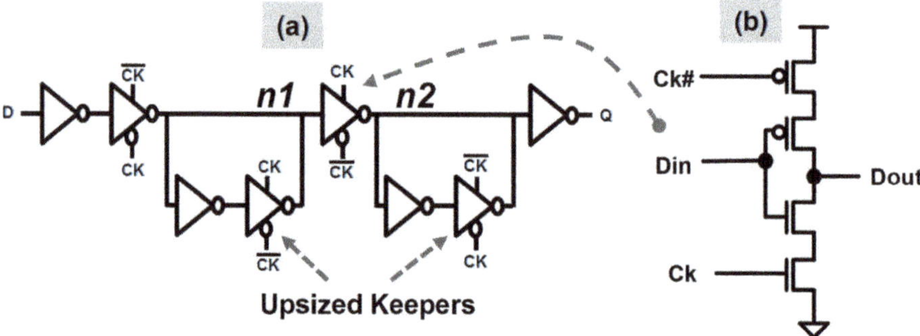

Figure 4. Low-voltage pass-gate free sequential circuit: (**a**) clocked-CMOS-style flip-flop implementation; (**b**) Clocked inverter logic gate.

Clock edges also degrade severely at low voltages, since local clock drivers inside flip-flops are small (Z_{MIN}) and prone to process variations. This can cause hold-time degradations and min-delay failures. The NTV-SIMD engine employs vector flip-flops, where the clock outputs of local drivers are shared across the flip-flops. As shown in Figure 5a, the drivers on nodes C, C# and Cd are shared across multiple flip-flops. This helps average the impact of device parameter variations. In addition, vector flip-flops enable lower clock power since they present a reduced input capacitance to the clock tree drivers. Under worst-case variation, if one of the local clock inverters becomes weak, the other shared clock inverter will compensate for the reduced drive strength. Vector flip-flop simulations (22-nm) across two adjacent cells with shared local min-sized clock inverters (Figure 5b) show better hold time violations at NTV and improved hold time margins by 175 mV. Stacked min-delay buffers limit variation-induced transistor speed up, further improving hold time margins at NTV by 7–30%.

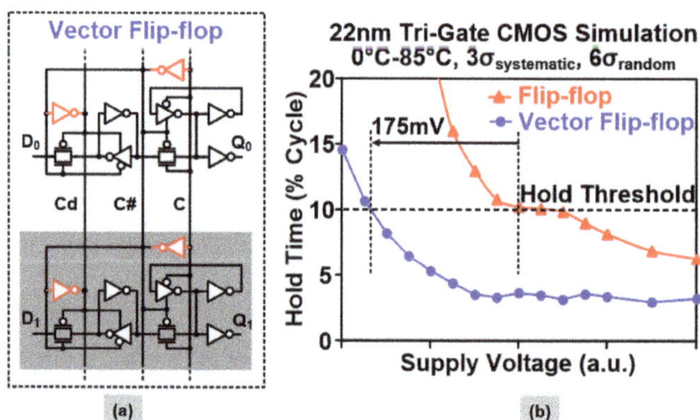

Figure 5. Averaging impact of device parameter variations at NTV: (**a**) vector flip-flop topology with shared clock input drivers; (**b**) Simulated hold time improvement in 22-nm node in the presence of random variations (6σ).

2.4. Level Shifter Circuit Optimizations

NTV designs, operating at low supply voltages require level shifters to communicate with circuits at the higher voltages (e.g., I/O). Similar to register file writes, conventional CVSL level shifters are inherently contention circuits. The need for wide range, ultra-low voltage level shifter to a high supply voltage further exacerbates this contention. The ultra-low voltage split-output, or ULVS, level shifter decouples the CVSL stage from the output driver stage and interrupts the contention devices, thus improving V_{min} by 125 mV (Figure 6). Full interruption of contention devices occurs for voltages $V_{in} \geq V_{out}$, while for voltages $V_{in} < V_{out}$ the contention devices are only partially interrupted, but still is beneficial at low voltages. For equal fan-in and fan-out, the ULVS level shifter weakens contention devices, thereby reducing power by 25% to 32%.

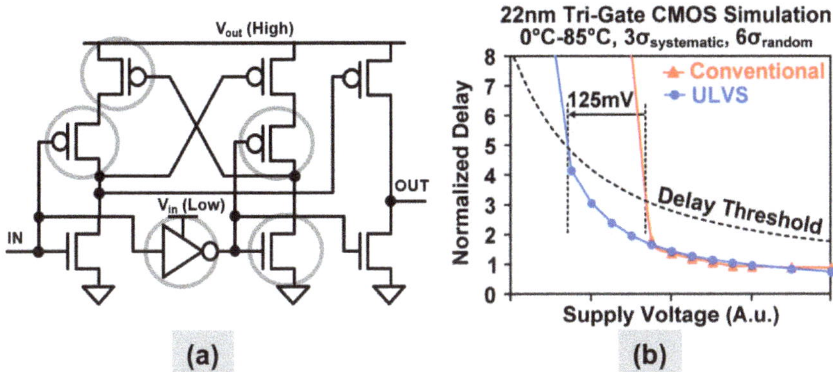

Figure 6. Ultra-low voltage split-output (ULVS) level shifter: (**a**) Circuit diagram with critical devices circled; (**b**) Simulated 22-nm V_{min} benefit of 125mV node in the presence of random variations (6σ).

Figure 7 summarizes improvements achieved from applying multiple circuit techniques for both the register file and logic circuits across 0–85 °C in the 22-nm SIMD engine. The static register file read circuits and shared P/N DETG write SRAM bit-cells improve overall register file V_{min} by 250 mV. Shared gates, ULVS level shifters, and vector flip-flops improve overall logic V_{min} by 150 mV.

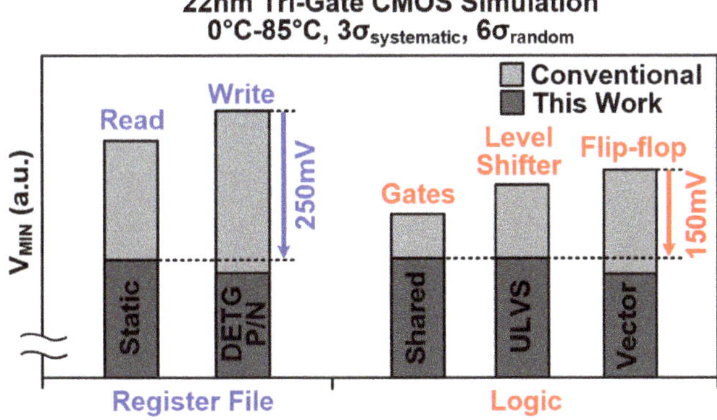

Figure 7. Simulated 22nm SIMD engine register file and logic optimization benefits across 0–85 °C, 3σ systematic, 6σ random variation.

3. Architecture Driven NTV Resilient NoC Fabrics

Architectural techniques can help regain some of the performance loss from engaging aggressive V_{DD} reduction. The limits to NTC-based parallelism to reclaim performance have been discussed in [18]. Dynamic adaptation techniques have been shown to monitor the available timing margin and guard bands in the design and dynamically modulate the voltage/frequency (V/F), thus preventing occurrence of timing errors [19]. Architecture-assisted resilient techniques, on the other hand, are more aggressive with the V/F push. In this case, the errors are allowed to happen, they are detected and then corrected using appropriate replay mechanisms.

Replica path-based methods such as tunable replica circuits (TRC) have been proposed [20] for error detection in flip-flop based static CMOS logic blocks. In this approach, a set of replica circuits are calibrated to match the critical path pipeline stage delay and timing errors are detected by double-sampling the TRC outputs. The key requirement is that the TRC must always fail before the critical path fails. The TRC is an area-efficient and non-intrusive technique, but it cannot leverage the probabilities of critical path activation, multiple simultaneous switching at inputs of complex gates, or worst case coupling from adjacent signal lines. An alternative in-situ approach for timing error detection uses error detection sequentials (EDS) in the critical paths of the pipeline stage. Timing errors are detected by a double-sampling mechanism using a flip-flop and a latch (Figure 8b) [21]. Errors are corrected by performing a replay operation at higher V or lower F. The V/F can also be adapted by monitoring the error rate and accounting for error recovery overheads.

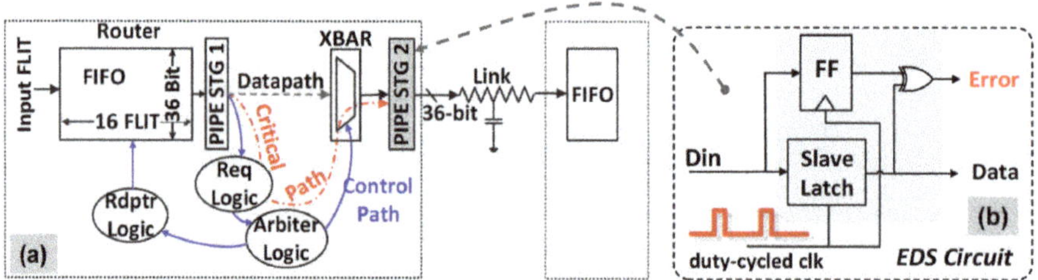

Figure 8. NTV-NoC: (**a**) Two-stage router data path and control logic indicating critical timing path; (**b**) Pipe stage 2 is enhanced with EDS circuit to detect failures in critical timing paths down to NTV.

NoCs have rapidly become the accepted method for connecting a large number of on-chip components. Packet-switched routers are key building blocks of NoCs [13]. Margins for operating V/F used to guarantee error-free operation limit achievable energy efficiency and performance at V_{OPT}. While error-correction codes (ECC) have been previously used to mitigate transient failures in routers [22], the associated performance and energy overheads can be significant for detection and correction of multi-bit failures. Timing error detection using EDS has been used for processor pipelines with minimal overhead [21]. An NTV router, designed in a 22-nm node and enhanced with EDS and a FLIT replay scheme, provides resilience to multi-bit timing failures for on-die communication. The goal is to evaluate the performance and energy benefits of single-error correction double-error detection (SECDED) ECC method over an EDS-based approach, from nominal V_{DD} down to NTV.

Resilient Router Architecture and Design

The 6-port packet-switched router in the 2 × 2 2-D mesh NoC fabric communicates with the traffic generator (TG) via two local ports and with neighboring routers using four bidirectional, 36-bit 1.5 mm long on-die links (Figure 1c). Inbound router FLITs are buffered in a 16-entry 36-bit wide FIFO (Figure 8a). The most critical timing path in the router consists of request generation, lane and port arbitration, FIFO read, followed by a fully non-blocking crossbar (XBAR) traversal. Any failure in this timing path is detected by the EDS circuit (Figure 8b) embedded in the output pipe stage (STG 2). The two-cycle EDS enhanced router can be run in two modes, with and without error detection. The TG contains SECDED logic which appends or retrieves nine ECC bits from a packet's tail FLIT, thus allowing end-to-end detection and correction of errors in the payload. A programmable noise injector [21] is introduced at each node on V_{NoC} supply to induce noise events during packet transmission.

The router control logic recovers from timing failures by saving critical states for the last *two* FLIT transmissions (Figure 9a). In the event of a timing failure, the *Error* signal generated by the EDS circuit in STG 2 is captured along with the erroneous FLIT in the recipient's FIFO, modified to accommodate an additional error bit as shown. Forward error correction is achieved by qualifying the FIFO output with the *Error* flag. In the router with the timing failure, the *Error* signal is latched to mitigate metastability. This synchronized *Error* flag is then used to roll-back the arbiters and FIFO read pointers to the previous functionally correct state. The current FLIT is again forwarded as part of replay. Error synchronization and roll-back incur two clock cycles of delay between an error event and successful recovery (Figure 9b). To avoid min-delay failures at STG 2, a clock with scan-tunable duty cycle control is implemented for the EDS latches. Additional min-delay buffers are inserted in the crossbar data path for added hold margin at a 2.4% area cost. In addition, the resilient router incurs the following overheads: (a) About 2.5% of router sequentials are converted to EDS; (b) Enabling replay causes a 10.5% increase in sequential count with 1.6% area overhead; and (c) The power overhead for the entire router is 8.7% with a 2.8% area cost.

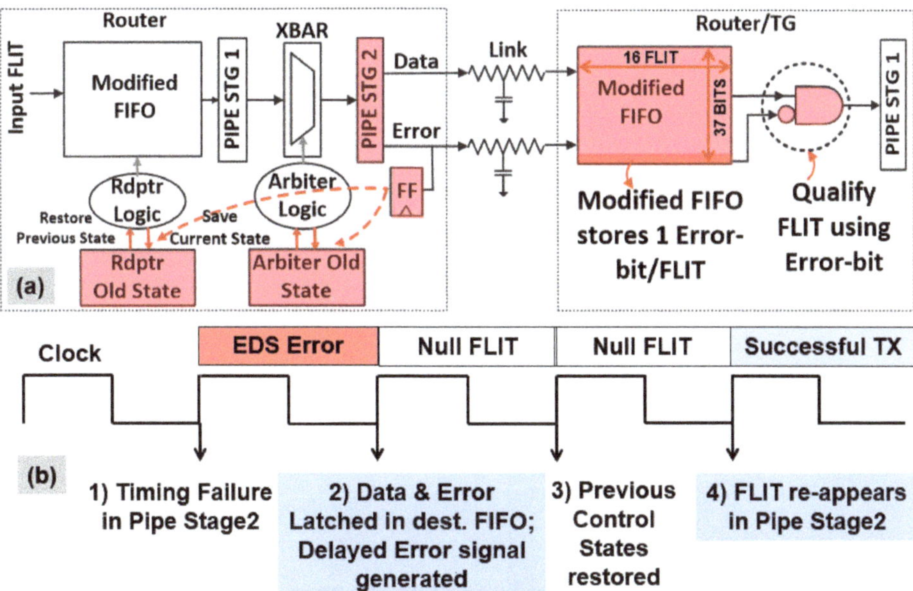

Figure 9. Internal router architecture: (**a**) Modifications to enable FLIT replay and forward error correction; (**b**) Clock cycle diagram showing stages of timing failure detection, replay and recovery.

4. Designing for Wide-Dynamic Range: Tools, Flows and Methodologies

Device optimizations need to work in concert with automated CAD design flows for optimal results. The 14-nm NTV-WSN design uses HP, standard-performance (SP), ULP, and thick-gate (TG)—all four transistor families in 14-nm second-generation tri-gate SoC platform technology [14]. To minimize variation induced skews, the clock distribution is completely designed using HP devices. The lower threshold voltage (V_T) of the HP devices allows improved delay predictability on the clock paths at NTV. SP devices are used for 100% of logic cells to achieve sufficient speeds during active mode of operation, with memory using ULP transistors for low standby power. The bidirectional CMOS IO circuits are designed using high voltage (1.8 V) TG transistors.

The optimized cell library for wide operational range is characterized at 0.5 V, 0.75 V and 1.05 V V_{DD} corners for design synthesis and timing convergence and are optimized for robust and reliable ultra-low voltage operation. Statistical static timing analysis (SSTA) is employed—a method which replaces the normal deterministic timing of gates and interconnects with probability distributions and provides a distribution of possible circuit outcomes [23,24]. As discussed in Section 2.2, variation-aware SSTA study is performed on the standard cell library to eliminate the circuits which exhibit DC failures or extreme delay degradation due to reduced transistor on/off current ratios and increased sensitivity to process variations. As a result, the standard cell library was conservatively constrained for use in the NTV optimized designs.

Achieving the performance targets across the entire voltage range is challenging since critical path characteristics change considerably due to non-linear scaling of device delay and a disproportionate scaling of device versus interconnect (wire) delay. It is critical to identify an optimal design point such that the targeted power and performance are achieved at a given corner without a significant compromise at the other corner. Synthesis corner evaluations for the NTV-CPU (Figure 10a) suggest that 0.5 V, 80 MHz synthesis achieves the target frequency at both 0.5 V (80 MHz) and 1.05 V (650 MHz). In comparison, it is observed that 1.05 V synthesis does not sufficiently size up the device dominated data paths which become critical at lower voltages, resulting in 40% lower performance at 0.5 V. Although 1.05 V synthesis achieves lower leakage and better design area, the 0.5 V corner was selected for final design synthesis of the NTV prototypes, considering its low voltage performance benefits and promise for wide operational range. Performance, area and power metrics at the two extreme design corners in a 32-nm node are presented in Figure 10b. For subsequent NTV prototypes, a multi-corner design performance verification (PV) methodology that simultaneously co-optimizes timing slack across all the three performance corners was developed. This PV approach ensures that performance targets are met across the wide voltage operational range. The method accounts for non-linear scaling of device delays in the critical path versus interconnect delay scaling across wide V_{DD}. At low voltages, severe effects of process variations result in path delay uncertainties and may cause setup (max) or hold (min) violations. Setup violations can be corrected by frequency binning. However, hold violations can cause critical functional failures. The design timing convergence methodology is enhanced to consider the effect of random variations and provide enough variation-aware hold margin guard-bands for robust NTV operation.

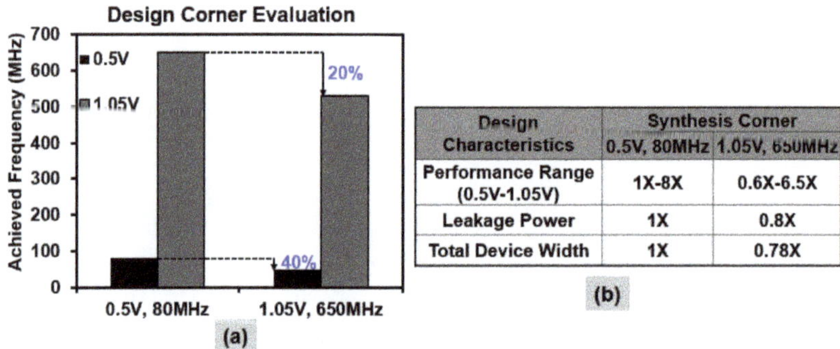

Design Characteristics	Synthesis Corner	
	0.5V, 80MHz	1.05V, 650MHz
Performance Range (0.5V-1.05V)	1X-8X	0.6X-6.5X
Leakage Power	1X	0.8X
Total Device Width	1X	0.78X

(b)

Figure 10. NTV-CPU: (a) Optimizations for wide range design convergence; (b) Design criteria varies widely at NTV (0.5 V) vs.1.05 V corner.

NTV Clocking Architecture

A calibrated ring oscillator (CRO) serves as a low-power on-chip high-frequency (MHz) clock source for the 14-nm NTV-MCU. The CRO is a frequency-locked loop (Figure 11a) that uses an RTC as a reference to generate a MHz clock output. Internally, the CRO tracks the frequency of oscillation from a ring oscillator and generates a delay code that adjusts the oscillation frequency to closely match the target frequency based on the reference clock. The CRO can operate in (1) closed-loop mode, where it accurately tracks the target frequency, as well as in (2) open loop mode at ultralow voltages, producing clock with tens of KHz frequency, enough for always-on (AON) sensing operation on the MCU. Silicon characterization data for the CRO is presented in Figure 11b. The on-die CRO locks to a wide range of target frequencies from 1 V down to 0.4 V. The CRO dissipates 60 μW (450 mV) while generating a 16 MHz output to clock the MCU at V_{OPT}. In open-loop condition, the CRO is functional down to a deep sub-threshold voltage of 128 mV, dissipating 3.8 μW, while generating a 7-kHz clock output. The CRO achieves a measured clock period jitter of 4.6 ps at 400-MHz operation.

The low-V_{DD} global clock distribution network on the NTV-CPU (Figure 11c) is designed with low-V_T devices to minimize clock skew across logic and memory voltage domain crossings, across the entire operating voltage range, and considers the effect of random variations. The clock tree incorporates two-stage level shifters and programmable delay buffers in the clock path. The level shifters in the clock path track the delay in the data-path level shifters. In addition, programmable lookup table based delay buffers can be tuned to compensate for any inter-block skew variations. SSTA (6σ) variation analysis shows 50% skew reduction at 0.5 V from clock delay tuning (Figure 11d).

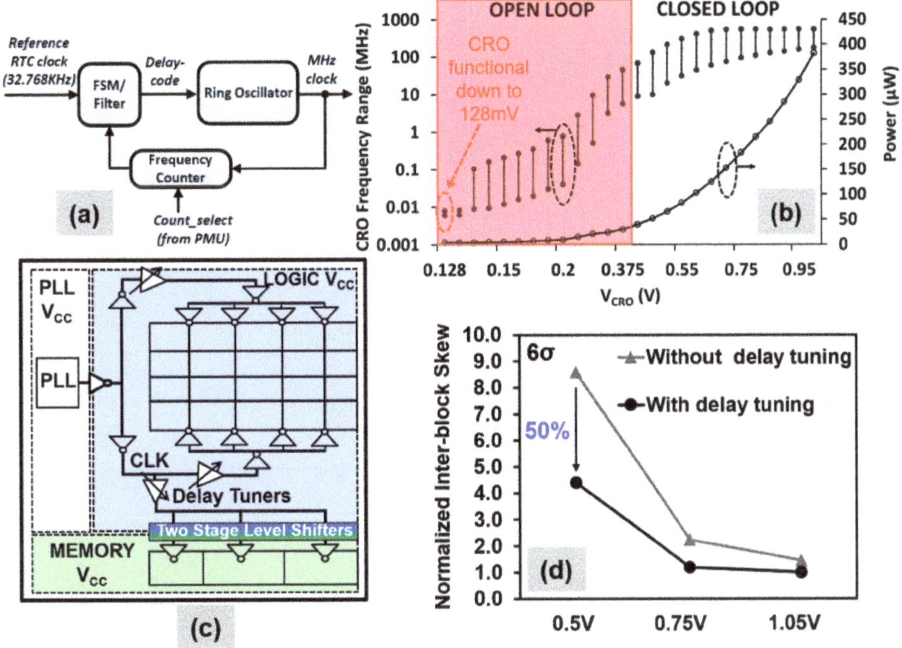

Figure 11. Multi-voltage global clock generation and distribution: (**a**) Calibrated Ring Oscillator (CRO) used in 14-nm NTV-MCU; (**b**) CRO operating range in both open/closed loops with μW power consumption; (**c**) The global clock distribution in the 32-nm NTV-CPU uses multi-stage level shifters; (**d**) Simulated clock skew reduction benefit from voltage specific delay tuning using look-up tables.

5. Key Results from Experimental NTV Prototypes

5.1. NTV-CPU Results

The NTV Processor is fabricated in a 32-nm CMOS process technology with nine layers of copper interconnect. Figure 12a shows the IA-32 die and core micrographs with a core area of 2 mm². Figure 12b shows the packaged IA processor and the solar cell (1 square inch area) used to power the core. The IA core is operational over a wide voltage range from 280 mV to 1.2 V. Figure 13 shows the measured total core power and maximum operational frequency (F_{max}) across the voltage range, measured while running the Pentium Built-In Self-Test (BIST) in a continuous loop mode. Starting at 1.2 V and 915 MHz, core voltage and performance scales down to 280 mV and 3 MHz, reducing total power consumption from 737 mW to a mere 2 mW. With a dual-V_{DD} design, memories stay at its measured V_{DD}-min of 0.55 V while allowing IA core logic to scale further down till 280 mV.

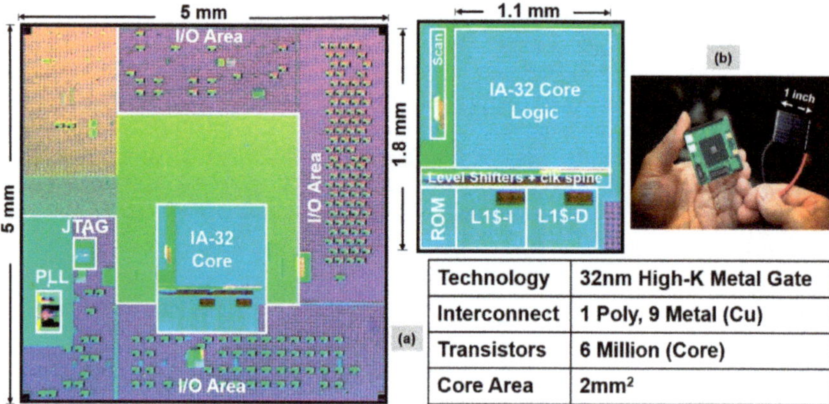

Technology	32nm High-K Metal Gate
Interconnect	1 Poly, 9 Metal (Cu)
Transistors	6 Million (Core)
Core Area	2mm²

Figure 12. NTV-CPU: (**a**) Full-Chip 32-nm die and core micrographs and characteristics; (**b**) Packaged IA-32 silicon and the small solar cell used to power the core.

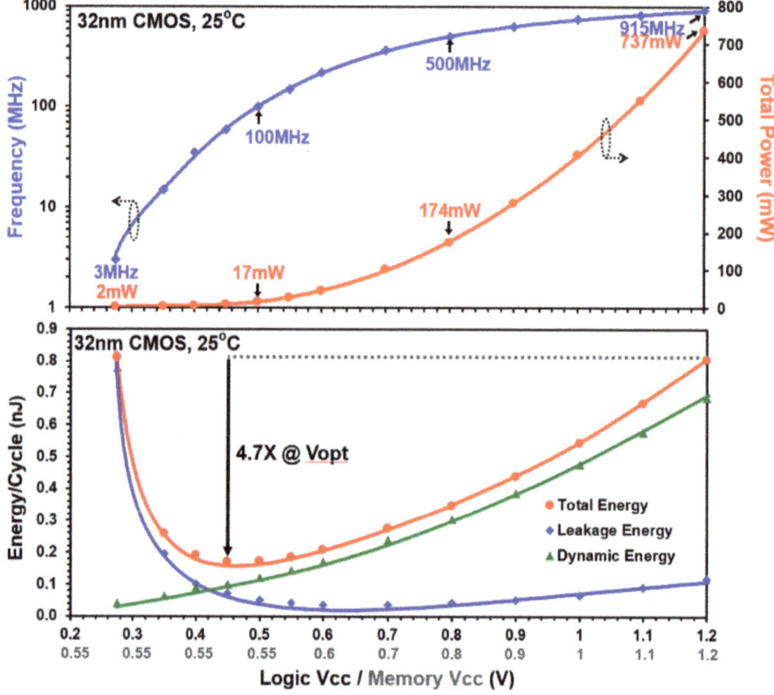

Figure 13. NTV-CPU measured power, performance and energy characteristics across wide V_{DD}.

Figure 13 also plots the measured total energy per cycle across the wide voltage range along with its dynamic and leakage components. Minimum energy operation is achieved at NTV, with the total energy reaching minima of 170 pJ/cycle at 450 mV (V_{OPT}), demonstrating 4.7× improvement in energy efficiency compared to the V_{DD}-max (1.2V) corner.

The pie-charts in Figure 14 shows a total core power breakup across super-threshold, near-threshold and sub-threshold regions. The contribution of logic dynamic power reduces drastically from 81% at V_{DD}-max to only 4% at V_{DD}-min (280 mV). Chip leakage power contribution as a proportion of total power starts increasing in the near-threshold voltage region and accounts for 42% of the total core power at V_{DD}-opt. At V_{DD}-min point, memories continue to stay at a higher V_{DD} than logic (550 mV), thus contributing 63% of the total core power.

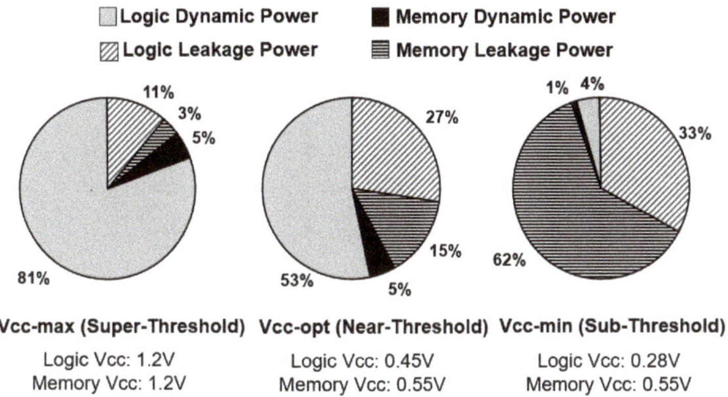

Figure 14. Measured NTV-CPU power breakdown across wide voltage range. Note that the memory supply scales down to 550mV, while the core logic operates well into the sub-threshold regime.

5.2. NTV-SIMD Engine Results

The SIMD permutation engine operates at a nominal supply voltage of 0.9 V and is implemented in a 22-nm tri-gate bulk CMOS technology featuring high-k metal-gate transistors and strained silicon technology. Figure 15 shows the die micrograph of the chip with a total compute die area of 0.048 mm^2. The permutation engine with 2-dimensional shuffle results in 36% to 63% fewer register file reads, writes, and permutes compared to a conventional 256b shuffle-based implementation. The SIMD engine contains 439,000 transistors.

Process	22nm Tri-Gate CMOS
Nominal Supply	0.9V
Register File Area	0.032mm^2
Permute Crossbar Area	0.016mm^2
Die Area	0.048mm^2
Number of Transistors	439K
Pad Count	30

Figure 15. NTV-SIMD vector engine die micrographs and characteristics.

Frequency and power measurements for the SIMD engine components are presented in Figure 16, obtained by sweeping the supply voltage from 280 mV to 1.1 V in a temperature-stabilized environment of 50 °C. Chip measurements show that the register file and crossbar operate from 3 GHz (1.1 V) down to 10 MHz (280 mV). The register file dissipates 227 mW (1.1 V) and 108 µW (280 mV) respectively, while the permute crossbar consumes 69 mW–19 µW over the same V_{DD} range. The maximum energy efficiency of 154 GOPS/W (1 OP = three 256b reads and one 256b write) is obtained at a supply voltage of 280 mV (V_{OPT}) and is 9× higher than the efficiency at nominal voltage. The 256b byte-wise any-to-any permute crossbar executes horizontal shuffle operations down to supply voltages of 240 mV.

Peak energy efficiency of 585 GOPS/W (1 OP = one 32-way 256b permutation) is achieved at a supply voltage of 260 mV, also with 9× better energy efficiency.

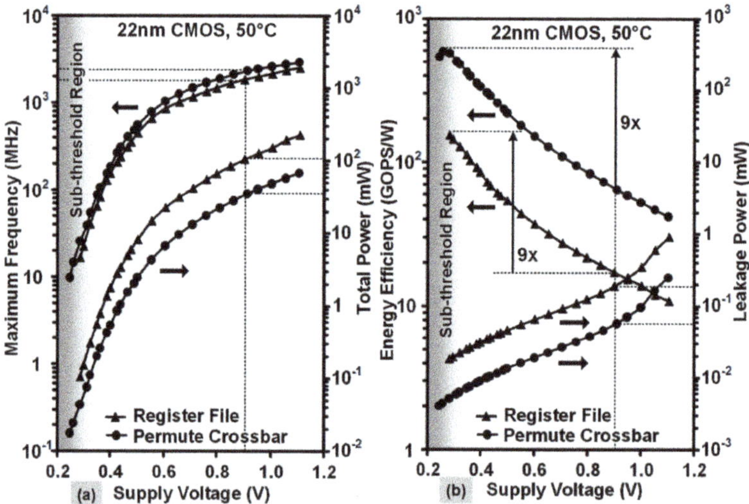

Figure 16. Measured NTV-SIMD engine: (**a**) Maximum frequency and power vs. V_{DD} (**b**) Energy efficiency vs. V_{DD}.

5.3. NTV-NoC Measurement Results and Learnings

The 2 × 2 2-D mesh-based resilient NoC prototype is fabricated in a 22 nm, 9-metal layer technology. Each router port features bidirectional, 36-bit wide, 1.5 mm long on-die links. The die area is 2.4 mm^2 with a NoC area of 0.927 mm^2 and a router area being 0.051 mm^2, as highlighted in the NoC die and NoC layout photographs (Figure 17a,c). There are approximately 31,400 cells in each router. The experimental setup and key design characteristics are shown in Figure 17b,d.

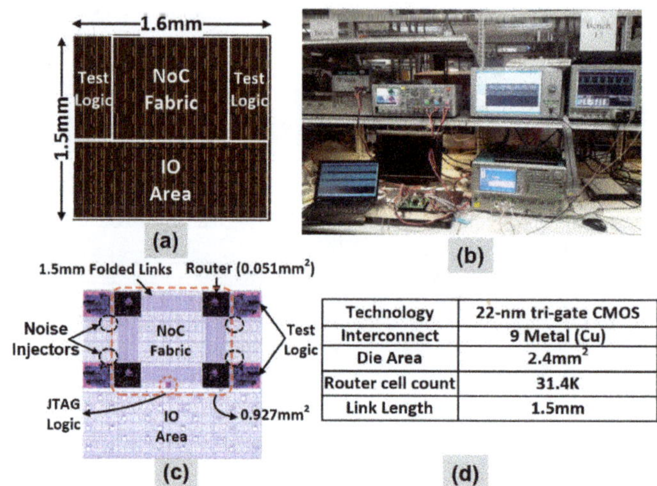

Technology	22-nm tri-gate CMOS
Interconnect	9 Metal (Cu)
Die Area	2.4mm^2
Router cell count	31.4K
Link Length	1.5mm

Figure 17. NTV-NoC design: (**a**) Die photograph with supply noise injectors; (**b**) Experiment setup; (**c**) NoC layout with key IP blocks identified and 1.5 mm folded links (**d**) Die characteristics.

Silicon measurements are performed at 25 °C for a representative NoC traffic pattern with FLIT injection at each router port every clock cycle at 10% data activity. The 2 × 2 NoC is functional over a wide operating range (Figure 18) with maximum frequency (F_{MAX}) of 1 GHz (0.85 V), 734 MHz (0.7 V), 151 MHz (400 mV), scaling down to 67 MHz (340 mV). A 3.3X improvement in energy-efficiency is achieved at a V_{OPT} of 400mV with an aggregate router bandwidth (BW) of 3.6 GB/s.

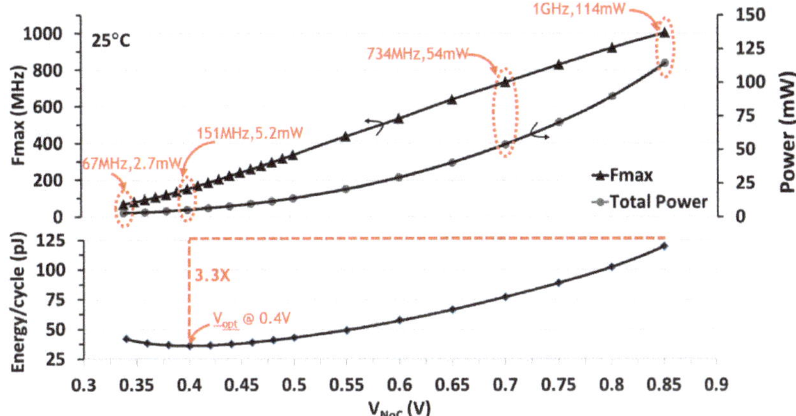

Figure 18. Measured 2 × 2 NoC power, performance and energy characteristics across wide voltage range with EDS and replay mechanisms disabled.

The measured NoC silicon logic analyzer trace (Figure 19) shows a supply noise-induced timing failure on the control bits of the packet header FLIT, followed by two cycles of bubble (null) FLITS and persistent retransmission (replay) of the FLIT until successful recovery. As shown, timing error synchronization and roll-back incurs a 2-cycle delay between an error event and successful recovery.

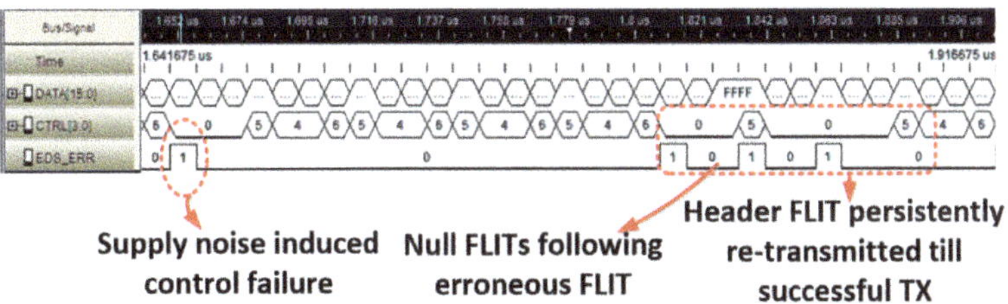

Figure 19. Silicon logic analyzer trace showing successful recovery of FLITs from timing failures.

Figure 20 plots the measured BW for the resilient router at 400 mV, in the presence of a 10% V_{NoC} droop induced by the on-die noise injectors. The number of erroneous FLIT increases exponentially with F_{CLK}. To account for such droop, a non-resilient router must operate with 28% (700 mV) and 63% (400 mV) F_{CLK} margins, respectively, thus limiting F_{MAX}. The resilient router reclaims these margins and offers near-ideal BW improvement until higher error rates and FLIT replay overheads limit overall BW gains. Past the point-of-first failure (PoFF), both control and data bits are corrupted. While ECC can identify data bit failures, control bit failures can invalidate the entire FLIT, rendering any ECC scheme ineffective. If control paths are designed with enough timing margins such that the control bits do not fail, the F_{CLK} gain from SECDED ECC is only 7% beyond PoFF, since several data bits fail simultaneously. In contrast, at 400 mV, the EDS scheme provides tolerance to multi-bit failures over a 9X wider F_{CLK} range, past PoFF. Compared to a conventional router implementation, the resilient router offers 28% higher bandwidth for 5.7% energy overhead at 700 mV and 63% higher bandwidth with 14.6% energy improvement at 400 mV.

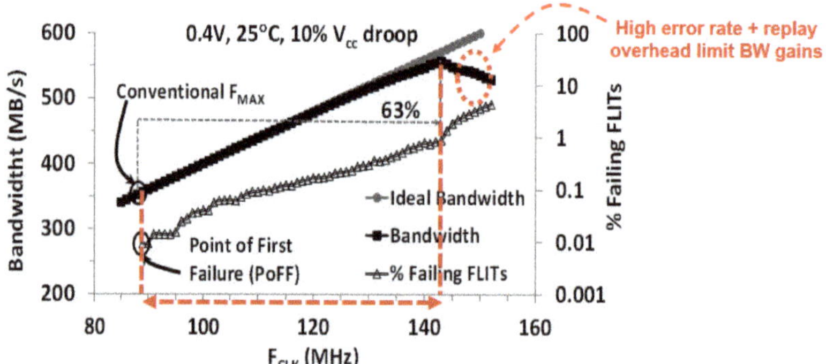

Figure 20. Improvement in resilient router bandwidth at V_{OPT} (400 mV) over a non-resilient version.

Resilience to Inverse Temperature Dependence Effects

As the supply voltage approaches V_T, elevated (lowered) silicon temperature results in increased (decreased) device currents. This phenomenon is generally known as Inverse Temperature Dependence (ITD) [25]. With process scaling and the introduction of high-κ/metal-gate, devices exhibit higher (negative) temperature coefficient along with weaker mobility temperature sensitivity [26]. This inverses the impact of temperature rise on delay, particularly as V_{DD} is lowered, where a small change in V_T results in a large current change—requiring large timing margins for NTV designs. As device and V_{DD} scaling exacerbates ITD, the need for characterizing and understanding ITD, and incorporating adaptive architectures becomes even more imperative. Measurements on the 22-nm NoC prototype indicate that ITD effects are observed at NTV, with router timing failures increasing as the die temperature decreases. Data in Figure 21 shows at 400 mV operation, a 30 °C temperature decrease (from 40 °C → 10 °C) causes the percentage of failing FLITs to rapidly increase. However, the resilient router recovers from transient timing failures due to EDS circuit error detection and the FLIT replay mechanism. This improves BW and F_{CLK} margins by 50% at 10 °C, when compared to a non-resilient router design.

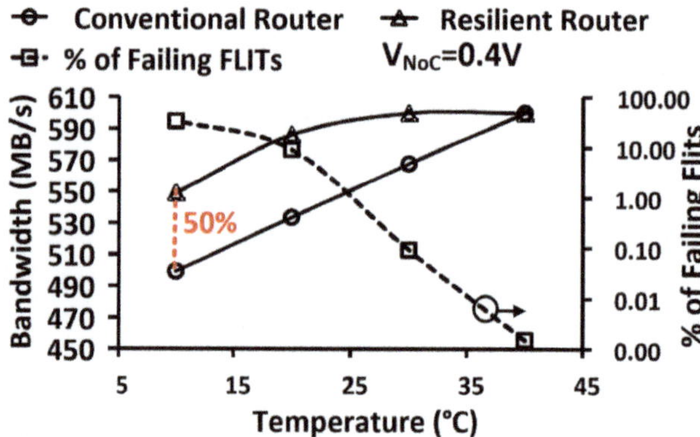

Figure 21. Measured NTV-NoC resilience to temperature variations and ITD effects at 400 mV. Timing failures in a conventional router increases at a faster rate over an EDS-enhanced one. The percentage of failing FLITS is indicated in the secondary y-axis.

5.4. NTV-MCU Measurement Results and WSN Operation

The MCU is fabricated using 14-nm tri-gate CMOS technology with nine metal interconnect layers (Figure 22). The MCU cell count is approximately 160 K and the die area is 0.79 mm^2 (0.56 mm × 1.42 mm). The surface-mount ball grid array (BGA) package has 24 pins with an area of 4.08 mm^2 (2.46 mm × 1.66 mm). The die photograph with key IP blocks identified and design characteristics are highlighted in Figure 23. The diminutive low-power MCU can serve as a key component for future autonomous, self-powered "smart dust" WSNs [27]—which can sense, compute, and wirelessly relay real-time information about the ambient.

Figure 22. The 32-b IA NTV MCU is packaged in a miniature 4.08-mm^2 (2.46 mm × 1.63 mm) 24-pin BGA substrate.

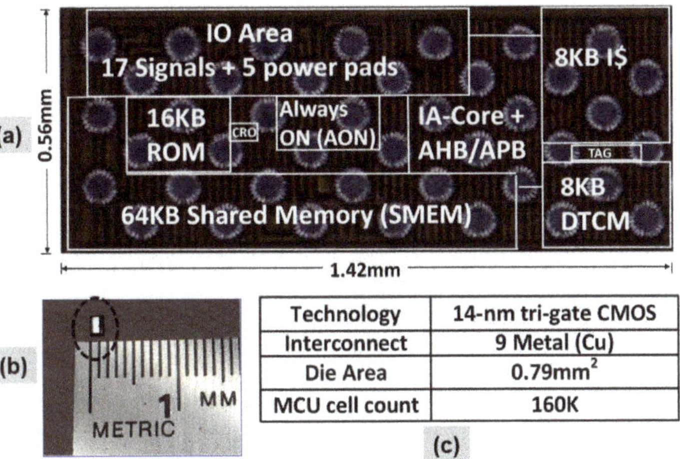

Technology	14-nm tri-gate CMOS
Interconnect	9 Metal (Cu)
Die Area	0.79mm^2
MCU cell count	160K

Figure 23. NTV-MCU 14-nm design: (**a**) Die photograph with key IP blocks identified. The die area is dominated by 64 KB of shared memory (SMEM); (**b**) Packaged 4 mm^2 die; (**c**) Die characteristics.

The IA MCU is functional over a wide operating range (Figure 24) from 297 MHz (1 V) scaling down to 0.5 MHz (308 mV) at 25 °C. While the entire MCU is functional down to 308 mV, SMEM functionality was validated down to 300 mV by independently writing and reading to it via the TAP debug interface. The ROM and the AHB logic are found to be functional down to 297 mV. With the MCU continuously executing a data encryption workload (AES-128), the minimum energy point is observed at 370 mV (V$_{OPT}$) at T = 25 °C. At V$_{OPT}$, the MCU operates at 3.5 MHz and dissipates 58 µW power, which translates to an energy-efficiency metric of 17.18 pJ/cycle. Compared to super-threshold operation at 1 V, NTV operation at V$_{OPT}$ achieves 4.8× improvement in energy efficiency.

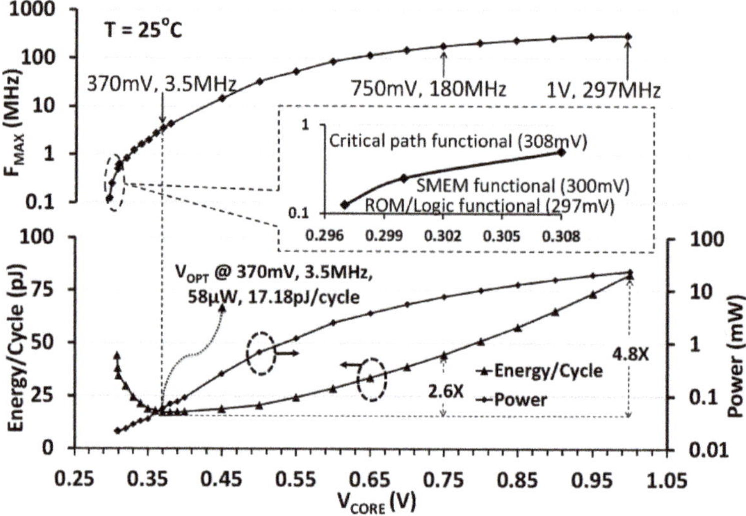

Figure 24. Measured 14-nm NTV-MCU power, performance, and energy efficiency across wide V_{DD}.

The MCU integrates 8 KB of Instruction cache (I$) and 8 KB data tightly coupled memory (DTCM). DTCM functions as a local scratch-pad memory, offering low latency (single cycle) and deterministic access, particularly valuable for data-intensive workloads. For typical WSN workloads with code footprint ~16 KB, MCU energy can be further improved by enabling I$ and DTCM. Enabling I$ and DTCM helps to exploit any code and data locality present in the application, thereby reducing the active power consumed in AHB interconnect and large SMEM (64 KB) access. Our experiments show 40% energy improvement is achievable from enabling both I$ and DTCM.

The WSN incorporating the NTV CPU operates continuously using the energy harvested by a 1 cm^2 solar cell from indoor light (1000 lux), with sensor data transmitted over BLE radio. The measured WSN power profile in AOAS mode over a 4-min interval is shown in Figure 25. In the AOAS operating mode (with BLE advertising + sensor polling every four seconds), average power (P_{AVG}) for the entire WSN is 360 µW, with the MCU contributing 290 µW (13 MHz, 0.45 V). The MCU power further drops to 120 µW in deep sleep state. In the deep sleep state, the core (IA + AHB) and CRO domains are power gated. The AON logic is still powered-ON and driven by RTC clock.

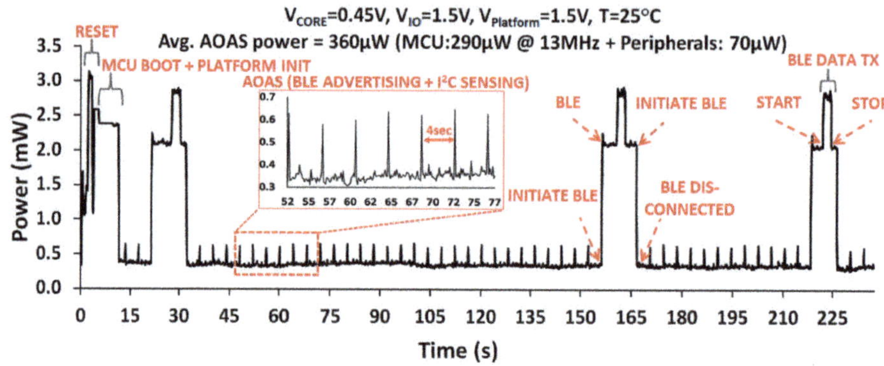

Figure 25. Measured WSN power profile in AOAS mode over a 4-min interval.

6. Conclusions and Future Work

NTV computing with wide dynamic operational range offers the flexibility to provide the performance on demand for a variety of workloads while minimizing energy consumption. The technology has the potential to permeate the entire range of computing—from ultra energy-efficient servers, personal and mobile computing to self-powered WSNs. It allows us to exploit the advantages of continued Moore's law to provide highest energy efficiency for throughput-oriented parallel workloads without compromising performance. The overheads of NTV design techniques in complex SoCs must be carefully balanced against impacts on power-performance at the higher end of the operating regime. Adaptive designs with in-situ monitoring circuitry can help detect and fix timing errors dynamically, but at an added cost. Four case-studies highlighting novel resilient architecture and circuit techniques, multi-voltage designs, and variation-aware design methodologies are presented for realizing robust NTV SoCs in scaled CMOS process nodes. In general, designs can tradeoff performance for reduced leakage power to realize better energy gains at NTV. The results demonstrate 3–9× energy benefits at NTV and the proposed design automation methodology can indeed help achieve greater energy reduction. As a future work project, we intend to build unified reliability models for NTC circuits and systems and validate the model against experimental data obtained across a wide voltage range.

Author Contributions: Conceptualization, S.V., S.P., S.H., A.A., R.K., J.T. and V.D.; methodology, S.V., S.H. and S.P.; software, S.P.; validation, S.V., S.P., A.A. and S.H.; investigation, S.P., A.A. and V.D.; resources, S.V.; data curation, S.P. and S.H.; writing—original draft preparation, S.V.; writing—review and editing, S.V., S.P., S.H., R.K. and V.D.; visualization, S.V.; supervision, S.V., R.K., J.T. and V.D.; project administration, S.V. All authors have read and agreed to the published version of the manuscript.

Funding: This research received no external funding.

Acknowledgments: The authors thank S. Jain, S. Khare, V. Honkote, M. Abbott, T. Majumder, P. Aseron, T. Nguyen, H. Kaul, M. Anders, M. Khellah, S. Mathew at Intel Labs, D. Mallik, and V. Grossnickle at Intel for technical contributions, encouragement, and support and the efforts of Intel's ATTD team with chip package design and assembly.

Conflicts of Interest: The authors declare no conflict of interest.

References

1. Ronald, G.; Dreslinski, M.; Wieckowski, D.; Blaauw, D.; Sylvester, D.; Mudge, T. Near-threshold computing: Reclaiming Moore's law through energy efficient integrated circuits. *Proc. IEEE* **2010**, *98*, 253–266.
2. De, V.; Vangal, S.; Krishnamurthy, R. Near Threshold Voltage (NTV) Computing. *IEEE Des. Test* **2016**, *34*, 1. [CrossRef]
3. Hanson, S.; Zhai, B.; Bernstein, K.; Blaauw, D.; Bryant, A.; Chang, L.; Das, K.K.; Haensch, W.; Nowak, E.J.; Sylvester, D.M. Ultra low-voltage, minimum energy CMOS. *IBM J. Res. Dev.* **2006**, *50*, 469–490. [CrossRef]
4. Jan, C.-H.; Agostinelli, M.; Buehler, M.; Chen, Z.-P.; Choi, S.-J.; Curello, G.; Deshpande, H.; Gannavaram, S.; Hafez, W.; Jalan, U.; et al. A 32nm SoC platform technology with 2nd generation high-k/metal gate transistors optimized for ultra low power, high performance, and high density product applications. *IEEE Int. Electron Devices Meet.* **2009**, 1–4. [CrossRef]
5. Schutz, J. A 3.3V 0.6um BiCMOS Superscalar Microprocessor. In Proceedings of the International Solid State Circuits Conference, Digest of Technical Papers, San Francisco, CA, USA, 16–18 February 1994; pp. 202–203.
6. Jain, S.; Khare, S.; Yada, S.; Ambili, V.; Salihundam, P.; Ramani, S.; Muthukumar, S.; Srinivasan, M.; Kumar, A.; Gb, S.K.; et al. A 280mV-to-1.2V wide-operating-range IA-32 processor in 32nm CMOS. In Proceedings of the 2012 IEEE International Solid-State Circuits Conference, Digest of Technical Papers, San Francisco, CA, USA, 19–23 February 2012; pp. 66–68.
7. Flachs, B.; Asano, S.; Dhong, S.; Hofstee, P.; Gervais, G.; Kim, R.; Le, T.; Liu, P.; Leenstra, J.; Liberty, J.; et al. A streaming processing unit for a CELL processor. In Proceedings of the ISSCC 2005 IEEE International Digest of Technical Papers Solid-State Circuits Conference, San Francisco, CA, USA, 10 February 2005; Volume 1, pp. 134–135. [CrossRef]

8. Hsu, S.; Agarwal, A.; Anders, M.; Mathew, S.; Kaul, H.; Sheikh, F.; Krishnamurthy, R. A 280mV-to-1.1V 256b reconfigurable SIMD vector permutation engine with 2-dimensional shuffle in 22nm CMOS. In Proceedings of the 2012 IEEE International Solid-State Circuits Conference, Institute of Electrical and Electronics Engineers, San Francisco, CA, USA, 19–23 February 2012; pp. 178–180.

9. Jan, C.-H.; Bhattacharya, U.; Brain, R.; Choi, S.-J.; Curello, G.; Gupta, G.; Hafez, W.; Jang, M.; Kang, M.; Komeyli, K.; et al. A 22nm SoC platform technology featuring 3-D tri-gate and high-k/metal gate, optimized for ultra low power, high performance and high density SoC applications. In Proceedings of the 2012 International Electron Devices Meeting, San Francisco, CA, USA, 10–13 December 2012; pp. 311–314. [CrossRef]

10. Paul, S.; Abbott, M.; Kishinevsky, E.; Aseron, P.; Vangal, S.; De, V.; Taylor, G. A 3.6 GB/s 1.3 mW 400 mV 0.051 mm^2 near-threshold voltage resilient router in 22-nm tri-gate CMOS. In Proceedings of the VLSI Circuits Symposium Digest of Technical Papers, Kyoto, Japan, 12–14 June 2013; pp. C30–C31.

11. Paul, S.; Honkote, V.; Kim, R.; Majumder, T.; Aseron, P.; Grossnickle, V.; Sankman, R.; Mallik, D.; Jain, S.; Vangal, S.; et al. An energy harvesting wireless sensor node for IoT systems featuring a near-threshold voltage IA-32 microcontroller in 14nm tri-gate CMOS. In Proceedings of the 2016 IEEE Symposium on VLSI Circuits (VLSI-Circuits), Honolulu, HI, USA, 15–17 June 2016; pp. 1–2. [CrossRef]

12. Vangal, S.R.; Howard, J.; Ruhl, G.; Dighe, S.; Wilson, H.; Tschanz, J.; Finan, D.; Singh, A.P.; Jacob, T.; Jain, S.; et al. An 80-Tile Sub-100-W TeraFLOPS Processor in 65-nm CMOS. *IEEE J. Solid-State Circuits* **2008**, *43*, 29–41. [CrossRef]

13. Vangal, S.; Singh, A.P.; Howard, J.; Dighe, S.; Borkar, N.; Alvandpour, A. A 5.1GHz 0.34mm2 Router for Network-on-Chip Applications. In Proceedings of the 2007 IEEE Symposium on VLSI Circuits, Kyoto, Japan, 14–16 June 2007; pp. 42–43. [CrossRef]

14. Jan, C.H.; Al-Amoody, F.; Chang, H.Y.; Chang, T.; Chen, Y.W.; Dias, N.; Hafez, W.; Ingerly, D.; Jang, M.; Karl, E.; et al. A 14 nm SoC platform technology featuring 2nd generation tri-gate transistors, 70 nm gate pitch, 52 nm metal pitch, and 0.0499 μm^2 SRAM cells, optimized for low power, high performance and high density SoC products. In Proceedings of the 2015 Symposium on VLSI Technology (VLSI Technology), Kyoto, Japan, 16–18 June 2015.

15. Intel Corporation. Intel Quark Processors. Available online: http://www.intel.com/content/www/us/en/embedded/products/quark/overview.html (accessed on 1 April 2020).

16. Raychowdhury, A.; Geuskens, B.; Kulkarni, J.P.; Tschanz, J.; Bowman, K.; Karnik, T.; Lu, S.-L.; De, V.; Khellah, M.M. PVT-and-aging adaptive wordline boosting for 8T SRAM power reduction. In Proceedings of the 2010 IEEE International Solid-State Circuits Conference (ISSCC), San Francisco, CA, USA, 7–11 February 2010; pp. 352–353. [CrossRef]

17. Kulkarni, J.P.; Geuskens, B.; Karnik, T.; Khellah, M.; Tschanz, J.; De, V. Capacitive-coupling wordline boosting with self-induced VCC collapse for write VMIN reduction in 22-nm 8T SRAM. In Proceedings of the 2012 IEEE International Solid-State Circuits Conference, San Francisco, CA, USA, 19–23 February 2012; pp. 234–236. [CrossRef]

18. Pinckney, N.; Sewell, K.; Dreslinski, R.G.; Fick, D.; Mudge, T.; Sylvester, D.; Blaauw, D. Assessing the performance limits of parallelized near-threshold computing. In Proceedings of the 49th Annual Design Automation Conference, Association for Computing Machinery (ACM), San Francisco, CA, USA, 3–7 June 2012; p. 1147.

19. Tschanz, J.; Lam, C.; Shuman, M.; Tokunaga, C.; Somasekhar, D.; Tang, S.; Finan, D.; Karnik, T.; Borkar, N.; Kurd, N.; et al. Adaptive Frequency and Biasing Techniques for Tolerance to Dynamic Temperature-Voltage Variations and Aging. In Proceedings of the 2007 IEEE International Solid-State Circuits Conference Digest of Technical Papers, San Francisco, CA, USA, 11–15 February 2007; pp. 292–604.

20. Tschanz, J.; Bowman, K.; Walstra, S.; Agostinelli, M.; Karnik, T.; De, V. Tunable replica circuits and adaptive voltage-frequency techniques for dynamic voltage, temperature, and aging variation tolerance. In Proceedings of the 2009 Symposium on VLSI Circuits, Digest of Technical Papers, Kyoto, Japan, 16–18 June 2009; pp. 112–113.

21. Bowman, K.A.; Tschanz, J.W.; Kim, N.S.; Lee, J.C.; Wilkerson, C.B.; Lu, S.-L.L.; Karnik, T.; De, V. Energy-Efficient and Metastability-Immune Resilient Circuits for Dynamic Variation Tolerance. *IEEE J. Solid-State Circuits* **2008**, *44*, 49–63. [CrossRef]

22. Rossi, D.; Metra, C.; Nieuwland, A.K.; Katoch, A. New ECC for crosstalk effect minimization. *IEEE Des. Test Comput.* **2005**, *22*, 340–348. [CrossRef]
23. Amin, C.S.; Menezes, N.; Killpack, K.; Dartu, F.; Choudhury, U.; Hakim, N.; Ismail, Y.I. Statistical static timing analysis. In Proceedings of the 42nd Design Automation Conference, Association for Computing Machinery (ACM), Anaheim, CA, USA, 13–17 June 2005; p. 652.
24. Singhee, A.; Singhal, S.; Rutenbar, R.A. Practical, fast Monte Carlo statistical static timing analysis: Why and how. In Proceedings of the 2008 IEEE/ACM International Conference on Computer-Aided Design, San Jose, CA, USA, 10–13 November 2008; pp. 190–195.
25. Cho, M.; Khellah, M.; Chae, K.; Ahmed, K.; Tschanz, J.; Mukhopadhyay, S. Characterization of Inverse Temperature Dependence in logic circuits. In Proceedings of the IEEE 2012 Custom Integrated Circuits Conference, San Jose, CA, USA, 9–12 September 2012; pp. 1–4.
26. Han, S.; Guo, D.; Wang, X.; Mocuta, A.C.; Henson, W.K.; Rim, K. Reverse Temperature Dependence of Circuit Performance in High-k/Metal-Gate Technology. *IEEE Electron Device Lett.* **2009**, *30*, 1344–1346.
27. Warneke, B.; Last, M.; Liebowitz, B.; Pister, K. Smart Dust: Communicating with a cubic-millimeter computer. *Computer* **2001**, *34*, 44–51. [CrossRef]

Journal of
*Low Power Electronics
and Applications*

Article

Cross-Layer Reliability, Energy Efficiency, and Performance Optimization of Near-Threshold Data Paths

Mehdi Tahoori * and Mohammad Saber Golanbari

Dependable Nano Computing, Karlsruhe Institute of Technology, 76131 Karlsruhe, Germany; golanbari@gmail.com
* Correspondence: mehdi.tahoori@kit.edu; Tel.: +49-721-608-47778

Received: 6 September 2020; Accepted: 18 November 2020; Published: 3 December 2020

Abstract: Modern electronic devices are an indispensable part of our everyday life. A major enabler for such integration is the exponential increase of the computation capabilities as well as the drastic improvement in the energy efficiency over the last 50 years, commonly known as Moore's law. In this regard, the demand for energy-efficient digital circuits, especially for application domains such as the Internet of Things (IoT), has faced an enormous growth. Since the power consumption of a circuit highly depends on the supply voltage, aggressive supply voltage scaling to the near-threshold voltage region, also known as Near-Threshold Computing (NTC), is an effective way of increasing the energy efficiency of a circuit by an order of magnitude. However, NTC comes with specific challenges with respect to performance and reliability, which mandates new sets of design techniques to fully harness its potential. While techniques merely focused at one abstraction level, in particular circuit-level design, can have limited benefits, cross-layer approaches result in far better optimizations. This paper presents instruction multi-cycling and functional unit partitioning methods to improve energy efficiency and resiliency of functional units. The proposed methods significantly improve the circuit timing, and at the same time considerably limit leakage energy, by employing a combination of cross-layer techniques based on circuit redesign and code replacement techniques. Simulation results show that the proposed methods improve performance and energy efficiency of an Arithmetic Logic Unit by 19% and 43%, respectively. Furthermore, the improved performance of the optimized circuits can be traded to improving the reliability.

Keywords: reliability; Near-Threshold Computing; functional unit; energy efficiency; performance optimization; cross-layer optimization

---

1. Introduction

Since the advent of electronic digital computing, relentless technology scaling has enabled an exponential improvement in computation capability while decreasing the cost and power consumption. Gordon Moore predicted in 1965 that the number of transistors in integrated circuits would double every year to address the ever-increasing demand for higher computation power [1] (this prediction was adjusted afterward to reflect the real progress [2,3]). Such continuous growth in computation capability over more than five decades affected almost all aspects of human life, including but not limited to industry, business, health-care, government, and society, which effectively started the Information Age, and made digital computing circuits an inseparable part of our everyday life.

Moore's law has faced several technological challenges and has been slowed down in the past decade [2,4,5]; however, still more transistors can be integrated on every new technology generation down to a 3 nm node [6,7].

To avoid exponential growth in the power density, various parameters including supply voltage of digital circuits have been scaled according to Dennard's scaling law [8]. However, the supply voltage did not scale at the same pace for about one decade (since 2005–2006) due to technological challenges associated with nanometer-scale devices [9]. Since then, various architectural and design directions have been actively explored including many-core computation, parallel processing, and 3D integration to provide more computation power [5–7]. However, the main challenges caused by the end of Dennard's scaling are energy efficiency and power density, which cannot be addressed by such approaches [10–12]. Without proper addressing of the increasing power density due to technology scaling, it is not possible to utilize all the components of a chip at the same time due to overheating, a problem commonly known as Dark Silicon [11,13], which diminishes the benefits of scaling. Therefore, reducing the power density through improving the energy efficiency is pivotal for future digital circuits.

In fact, energy-efficient computing has already become a primary requirement in various application domains. At one end of the spectrum, the growing IoT applications, with an expected 20 billion connected devices by 2020 [14–16], are continually looking for more energy efficiency. These IoT devices are expected to operate on limited energy sources such as batteries or energy harvesting sources. High-performance servers and data centers, at the other end, are responsible for a significant amount of consumed electricity worldwide (about 1.4% of the consumed electricity worldwide in 2011, and growing in much faster speed compared to other electricity consumers) [17]. A large portion of the cost of data centers is directly or indirectly due to the energy consumption of digital circuits [17]; hence, it is necessary to improve the energy efficiency of high-performance computing as well.

Power consumption of digital circuits has two components: dynamic power consumption, due to circuit activity and computation, and leakage power consumption, due to slight leakage current of transistors. Reducing the power density can be achieved through various approaches such as optimizing design techniques, employing power management strategies, improving the technology, and scaling supply voltage [18]. Each of these approaches aims at reducing dynamic power, leakage power, or both. In this regard, optimizing Instruction Set Architecture (ISA), scheduling methods, pipeline design, and synthesis methodologies have already enabled vast improvements in the energy efficiency [19].

Techniques such as clock-gating and power-gating are widely used in existing digital circuits to cut down dynamic and leakage power of the idle components [20]. Dynamic Voltage and Frequency Scaling (DVFS) promotes supply voltage and speed adaptation depending on the workload to reduce power consumption under low workload [21,22].

Aggressive supply voltage reduction down to the sub-threshold region is known as an important instrument, which reduces power consumption by several orders of magnitude and improves energy efficiency [12,18,23–25]. Intel has demonstrated ultra-low power characteristics using such aggressive voltage scaling by its experimental IA32 processor, which can run Windows and Linux on a small solar panel [26]. In addition to extensive power reduction, voltage scaling degrades circuit speed significantly. Many applications have performance constraints that prevent them from operating in the sub-threshold region.

By operating digital circuits at supply voltages close to the threshold voltage of the transistors, which is known as NTC, it is still possible to gain very high energy efficiency and achieve enough performance for many applications in the IoT domain [27,28]. Additionally, many-core computation based on NTC is an attractive approach to improve energy efficiency while satisfying the computational demands for highly parallel workloads of data centers [10,29–31]. Therefore, NTC is considered an attractive paradigm for improving the energy efficiency in modern technology nodes [32], if the associated reliability challenges are addressed. These reliability challenges are typically caused by enormous sensitivity of NTC circuits to variability sources and complicate NTC circuit design and operation.

Along with the enormous energy benefits, NTC comes with a variety of design challenges. The most obvious one is the performance reduction of $10\times$ compared to the super-threshold domain, which may limit the applicability of NTC [27]. In addition, the escalated sensitivity to variability (such as process and voltage variation) at reduced supply voltages forces designers to add very conservative and expensive timing margins to achieve acceptable yield and reliability [33].

Moreover, in the near-threshold region leakage energy becomes comparable to dynamic energy, thus approaches to minimize leakage are of uttermost importance for NTC designs. Hence, although conventional designs can technically operate in the NTC domain, due to these challenges, new design paradigms have to be developed for NTC to harness its full potential.

Data paths are core components of any processing element such as processor cores and accelerators. A data path consists of various functional units, such as Arithmetic Logic Unit (ALU) and Floating Point Unit (FPU), the timing and power consumption characteristics of which significantly impact the overall performance of the processor. Therefore, optimizing the reliability, energy efficiency, and performance of data paths is of decisive importance.

This paper presents two cross-layer functional unit optimization opportunities based on (1) instruction multi-cycling and (2) functional unit partitioning, which improve energy-efficiency, reliability, and performance. We evaluate our methodology by using an ALU implemented for Alpha ISA. Our results show that the proposed approach can effectively improve energy efficiency and reliability. For example, instruction multi-cycling effectively removes the extra timing slacks and improves the energy efficiency of a circuit by up to 34%. Furthermore, the functional unit partitioning approach can improve the energy efficiency of an ALU by 43.4% while having positive impacts on the reliability and performance as well.

The rest of this paper is organized as follows. Section 2 provides the background on near threshold computing. Section 3 reviews the state-of-the-art and their shortcomings. The optimization approaches based on instruction multi-cycling and functional unit partitioning are presented in Section 4, while Section 5 discusses the optimization results. Finally, Section 6 concludes the paper.

2. Near-Threshold Computing

Various methodologies at different abstraction-levels have been proposed and employed [18–22,34–36] to improve the energy efficiency and overcome the power wall challenge caused by the end of Dennard's scaling [9–12]. Aggressive supply voltage scaling down to the sub-threshold region [12,18,23–25] is also presented as an effective way to reduce the power consumption by several orders of magnitude; however, the speed of digital circuits at that supply voltage range is poor. In the following, we present a promising approach towards supply voltage scaling, called NTC, which also retains enough performance for many applications.

Near-Threshold Computing is a paradigm in which the supply voltage is reduced close to the threshold voltage of transistors to gain large energy efficiency while retaining enough performance for many applications. However, there are various challenges towards NTC mainly in the area of reliability and energy efficiency, which can nullify the benefits of NTC if not addressed correctly. This paper focuses on the NTC and its challenges.

From a circuit-level point of view, the total power consumption of a circuit can be calculated as in Equation (1):

$$P_{total} = P_{dyn,sw} + P_{dyn,sc} + P_{leak}, \tag{1}$$

where $P_{dyn,sw}$, $P_{dyn,sc}$, and P_{leak} are dynamic switching power, dynamic short-circuit power, and leakage power, respectively.

In the nominal supply voltage, the $P_{dyn,sw}$ is the dominant power consumption, which is quadratically dependent on the supply voltage, as shown in Figure 1a. Therefore, reducing the supply voltage can quadratically reduce the overall power consumption. In the same supply voltage regime, the speed of a circuit is linearly dependent on the supply voltage.

The energy consumption is calculated according to Equation (2):

$$E_{total} = P_{total} \times T_{clk} = E_{dyn,sw} + E_{dyn,sc} + E_{leak} \tag{2}$$

Therefore, by slightly reducing the supply voltage from the nominal voltage, it is possible to reduce the energy consumption, linearly. As shown in Figure 1b, this energy improvement holds as long as the rate of total power consumption reduction is higher than the rate of speed degradation. When V_{dd} goes below V_{th}, the speed degradation becomes exponential because the current has an exponential relation with supply voltage. Therefore, energy consumption starts to increase due to higher rate of speed degradation. As a result, the supply voltage leading to minimum energy consumption is somewhere close to the threshold voltage of transistors. Such operating condition leading to the best energy efficiency, i.e., the lowest energy consumption, is commonly known as the Minimum Energy Point (MEP) and is shown in Figure 1b for c499 circuit from ISCAS'85 benchmark circuits [37,38].

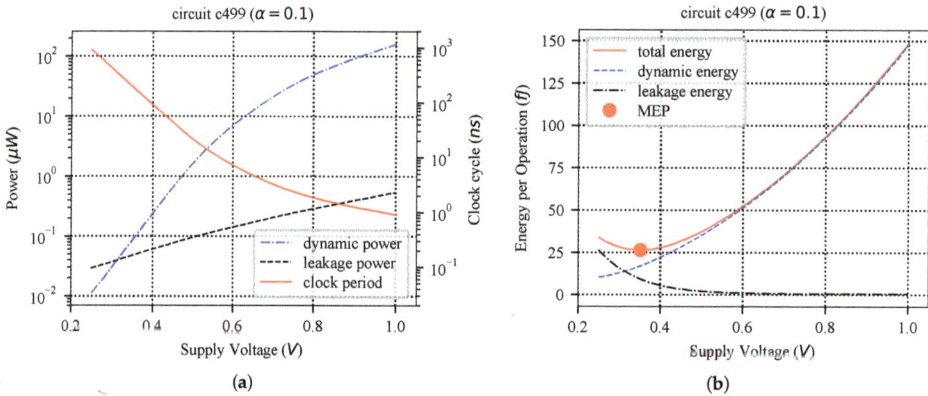

Figure 1. MEP exploration for circuit c499 from ISCAS'85 benchmark [37,38]. (**a**) Circuit power and clock period versus supply voltage (V_{dd}); (**b**) Minimum Energy Point (MEP). Minimum Energy Per operation is achieved where V_{dd} is close to V_{th}.

2.1. MOSFET Model in the Near-Threshold Voltage region

The conventional three-region long-channel MOSFET model as well as the alpha-power model [39] are piece-wise models with a discontinuity at the threshold voltage of transistors, which makes them inappropriate for NTC circuit analysis. However, it is possible to explain the characteristics of MOSFET based on continuous models such as EKV [40–42]. Accordingly, a simplified trans-regional model for digital NTC CMOS circuits is proposed in [43], which facilitates analytical analysis. Based on this model, the MOSFET on-current can be obtained based on the overdrive voltage V_{ov} as follows:

$$I_{ds,NTC} = I_x k_0 e^{k_1 \frac{V_{ov}}{nV_T} + k_2 \left(\frac{V_{ov}}{nV_T}\right)^2}, \qquad V_{ov} = V_{gs} - V_{th}, \tag{3}$$

where I_x depends on process parameters and transistor dimensions (W, L), whereas k_0, k_1, and k_2 are process independent fitting parameters [43]. In the above equation, it is assumed that $V_{ds} \gg V_T$; therefore, the term depending on V_{ds} is eliminated. However, it is possible to consider the threshold voltage dependency on V_{ds} and body-biasing. Based on this model, the propagation delay of a gate in the Near-Threshold Voltage (NTV) region is obtained as:

$$t_{p,NTC} = \frac{k_f C_L V_{dd}}{I_x k_0} e^{-k_1 \frac{V_{dt}}{n V_T} - k_2 \left(\frac{V_{dt}}{n V_T}\right)^2}, \qquad V_{dt} = V_{dd} - V_{th}. \tag{4}$$

Similarly, energy and power consumption of digital circuits can be calculated in the NTV region [43].

Process, Voltage, and Temperature Variation in NTC

The impacts of process, supply voltage, and temperature variations (PVT) are more pronounced in the NTV region [27,33,44]. Authors of [45] showed that the sensitivity of the drain-source current of a transistor (I_{ds}) to changes in V_{th} and V_{dd} increases by about $10\times$ when the supply voltage is reduced from the super-threshold region to the sub-threshold region. Equation (4) also demonstrates that propagation delay in the NTV region is exponentially dependent on supply and threshold voltages. Intel [33] reported that while the process and temperature variations cause 18% and 5% performance variation in the super-threshold region, their impacts aggravate to $2\times$ performance variation in the NTV region.

The power consumption of NTC circuits is orders of magnitude smaller than the super-threshold region. As a result, runtime supply voltage fluctuation caused by power consumption also decreases by the same scale, which makes it insignificant in NTC circuits, even considering the large sensitivity to fluctuations. Moreover, the temperature of NTC circuits is solely determined by the ambient temperature since the power dissipation is very small and has a negligible impact on circuit temperature fluctuation [46].

2.2. NTC Challenges

2.2.1. Reliability

Aggressive supply voltage scaling to the NTV region has benefits and drawbacks in terms of reliability. Reducing the supply voltage to the NTV region greatly reduces internal electric fields and current density compared to the super-threshold region, which helps to protect the circuits from some aging phenomena and their associated reliability problems.

Nevertheless, the exponential sensitivity to PVT in the NTV region severely affects the circuit behavior and may lead to large performance variation or reliability issues, in terms of functional failure or timing violations. As presented in [33], $\pm2\times$ performance variation is observed between different fabricated NTC cores only due to process variation, whereas this variation was limited to $\pm18\%$ at the nominal voltage. Techniques such as adaptive body biasing [27] and supply voltage scaling [47] are presented to address such performance fluctuation. Similarly, the maximum clock frequency of an NTC circuit may change by $\pm2\times$ due to 110 °C temperature fluctuation, while the impact of such temperature fluctuation on performance is only 5% at the nominal voltage [33] (the measurement is done for a 65 nm typical die). An unwanted performance fluctuation at gate-level may lead to timing violations. Setup-time violations can be addressed by increasing the clock period T_{clk} (i.e., reducing clock speed f), which inflicts performance and energy efficiency loss. Various timing error detection and correction methods have also been proposed to address setup-time issues [48–52], which may be costly as the number of timing errors increases rapidly in the NTV region. However, a hold-time violation cannot be fixed by changing the clock speed and leads to a functional failure. As an example, the number of hold-time violations increases by up to $16\times$ when operating in the NTV region compared to the nominal voltage [53]. In order to fix these timing violations, many buffers have to be inserted into the violating paths to delay the arrival times of such short paths. Therefore, the overhead of buffer insertion in the NTV region is significantly larger than that of the super-threshold region.

Reduced noise margin due to voltage scaling is a major challenge in storage component design for NTC. Conventional 6T SRAM memory cells cannot operate correctly in the NTV region without redesigning [54], and 8T or 10T SRAMs are preferred due to better noise margin and resiliency to variations [55–58]. In addition, due to low activity of memory cells, the optimum supply voltage may be considerably higher than core logic. Therefore, voltage level converters may be needed to interface memory and core cells, which brings additional complexity to routing and Power Delivery Network (PDN) design. Similarly, flip-flops have issues due to reduced noise margin and high sensitivity to variations [27,33]. Therefore, various flip-flop designs have been studied for low-voltage operation and are optimized for NTC [27,28,33,59–61]. The soft-error rate of storage components is dependent on the critical charge, which decreases by $5\times$ in the NTV region [62]. Therefore, more soft-errors are observed in the NTV region in both logic and memory components. Various methods have been proposed at different abstraction levels from device to architecture-level to address the faults caused by soft-errors for NTC [63–70].

2.2.2. Energy Efficiency

The energy efficiency of NTC circuits is highly dependent on design, process variation, and runtime parameters. Scaling down the supply voltage changes the ratio of dynamic and leakage power consumption significantly, causing a paradigm shift in design and optimization. Figure 2 presents the power consumption of an IA32 processor. The contribution of logic dynamic power decreases from 81% in the super-threshold region to 4% in the sub-threshold region, whereas the contribution of logic leakage power increases from 11% to 33% [26]. Therefore, leakage power reduction techniques at architecture-level (e.g., by power-gating schemes) down to device-level (e.g., by incorporating advanced technologies such as FinFET) are more rewarding in the NTV region compared to the super-threshold region, in which leakage power has a smaller contribution.

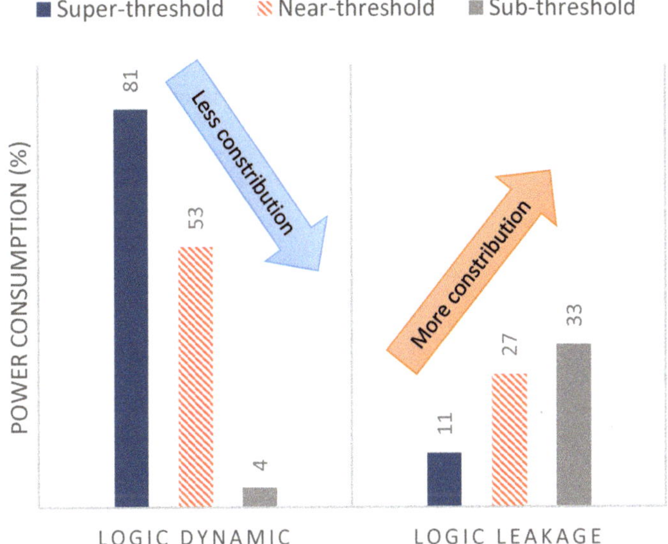

Figure 2. Power breakdown of the IA32 processor presented in [26] highlights a significant increase in the relative contribution of the leakage power of logic components to the total power consumption, when supply voltage is reduced towards near-threshold and sub-threshold regimes. The rest of the power consumption is due to the memory (8%, 20%, and 63% in the super-threshold, near-threshold, and sub-threshold region, respectively.)

The methods used for addressing the variation-induced reliability challenges affect the energy efficiency as well. Some of the methods commonly used in the super-threshold region for controlling the impact of variations are too expensive in the NTV region. As an example, adding a timing margin to compensate for variation-induced timing fluctuation is energy inefficient in the NTV region due to the extent of variations. Therefore, it is necessary to design the circuits to be more resilient against timing variation, for example, by variation-aware circuit synthesis.

The MEP of a circuit points to a specific supply voltage V_{dd}^{MEP} and the maximum speed f^{MEP} (inversely proportional to clock period $T_{clk}^{MEP} = 1/f^{MEP}$) at that supply voltage. A number of factors impact the MEP of a circuit including technology, internal activity, workload, and process and runtime variations. FinFET technology offers close to 60 mV/dec sub-threshold slope leading to better Ion/Ioff ratio, which is very useful in reducing the MEP and improving the energy efficiency. It has been shown that the MEP may move from a sub-threshold voltage region to super-threshold voltage region depending on the circuit structure and circuit internal switching activity [32]. For example, the contribution of leakage power to the total power, i.e., P_{leak}/P_{total}, is typically larger in SRAM arrays compared to the core logic. Therefore, the MEP of SRAM arrays is higher than core logic [32]. This also means that workloads that cause higher internal switching activity (high dynamic power), will reduce the MEP of a circuit. Process and runtime variations significantly contribute to MEP fluctuation. Therefore, design optimization and runtime tuning are also necessary to achieve high energy efficiency in the NTV region.

The average impact of process variation on V_{dd}^{MEP} of the benchmark circuits is displayed in Figure 3a. The shift in V_{dd}^{MEP} due to process variation is on average 66 mV for the benchmark circuits. This shift in V_{dd}^{MEP} may lead to significant performance variation and energy overheads. Moreover, the speed (T_{clk}^{MEP})

of some circuits is also highly affected by process variation. Figure 3b demonstrates that T_{clk}^{MEP} of a circuit may change by about one order of magnitude when it is under different process variation impacts.

The impact of temperature variation is shown in Figure 4, where the MEP is plotted for different temperatures ranging from -25 to 100 °C. The clock period corresponding to the MEP (T_{clk}^{MEP}) is also greatly affected by the change in the circuit temperature, with an exponential dependency. Comparing Figures 3 and 4 reveals that the impact of temperature could be much stronger than the impact of the process variation. This is also in line with the reported process and temperature variation sensitivities as in [27,28,33,47].

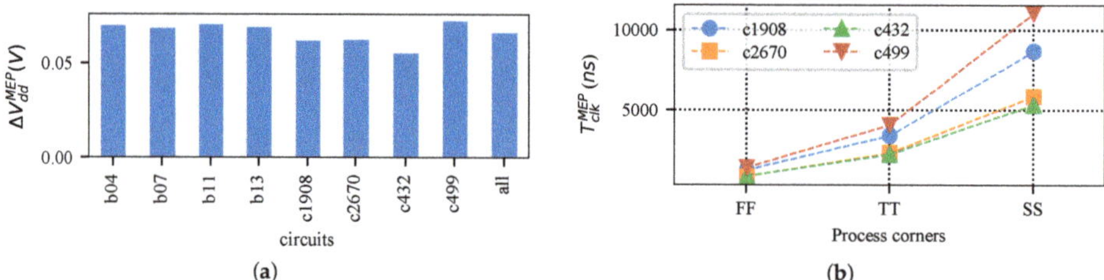

Figure 3. Process variation impact on the MEP of some ISCAS'85 benchmark circuits [71]. (**a**) Process variation can change V_{dd}^{MEP} on average by 66 mV. (**b**) Process variation can change T_{Clk}^{MEP} by up to an order of magnitude over different process corners.

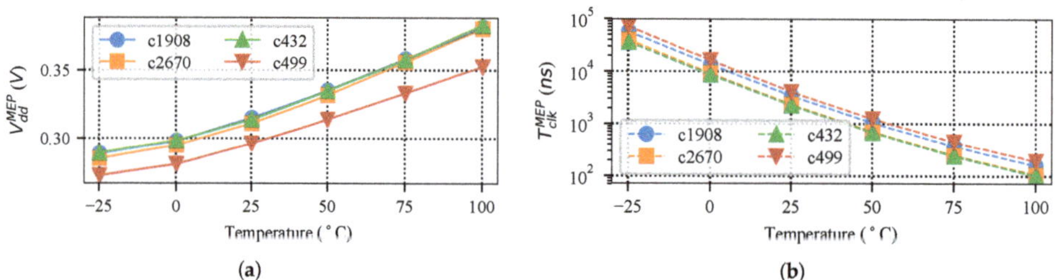

Figure 4. Temperature variation impact on the MEP of some ISCAS'85 benchmark circuits [71]. (**a**) Up to 80 mV variation in V_{dd}^{MEP} is observed over 125 °C temperature change. (**b**) More than 2 orders of magnitude variation in T_{clk}^{MEP} is observed over 125 °C temperature change.

Figure 5 shows the change in V_{dd}^{MEP} for circuit c499 of ISCAS'85 benchmark due to a change in the internal switching activity. The x-axis of the figure represents the ratio of the dynamic energy to the leakage energy of the circuit (Dynamic/Leakage), when the circuit is operating at the nominal supply voltage (V_{dd}^{NOM}). The Dynamic/Leakage value has an inverse relation with V_{dd}^{MEP} as demonstrated in Figure 5. According to this figure, if the input switching activity α changes from 0.01 to 0.1, the V_{dd}^{MEP} can vary from 0.49 V down to 0.35 V as Dynamic/Leakage increases. In a pure combinational circuit such as c499, the impact of workload variation could be very high as the dynamic power consumption P_{dyn} is dependent on the input switching activity α. However, in sequential circuits, a large portion of dynamic power consumption is related to the clock network. In such cases, P_{dyn} variation due to α fluctuation is typically less than in combinational circuits. Therefore, workload variation can also have a significant impact on energy efficiency, depending on the circuit architecture.

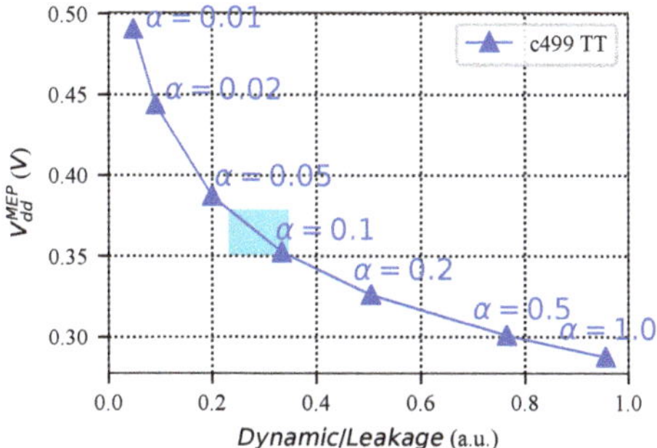

Figure 5. Workload variation impact on the MEP. The change in the V_{dd}^{MEP} versus the ratio of dynamic to leakage energy at nominal V_{dd}. The realistic change in the input switching activity α is displayed by the light blue box ($0.062 \leq \alpha \leq 0.108$) [71].

In summary, the energy efficiency of NTC circuits is highly affected by PVT. Design optimization methods can effectively improve energy efficiency by reducing the leakage power. Runtime tuning methods can adapt the circuit to operate at the MEP, which fluctuates at runtime due to temperature and workload fluctuation.

2.2.3. Wide-Voltage Resiliency

Some NTC circuits are required to operate at different supply voltage modes, due to runtime MEP tuning, ultra-low power requirements, or speed constraints (super-threshold). When a task with specific timing constraints executes on the NTC circuit, the supply voltage may be increased to the super-threshold region to meet the performance constraints. However, this brings reliability issues associated with super-threshold voltage, such as aging, into consideration as the circuit may operate in that mode for some time. For example, an IoT edge device may be assigned to process some data within a specific amount of time and send them to an IoT gateway. Therefore, it is necessary to take the aging issues into consideration if a device is expected to operate over a wide-voltage range as this may have deteriorating impact on the device operation when it is switched back to the NTV region or even in the super-threshold voltage region.

2.2.4. Other Challenges

Static Timing Analysis (STA) tools are typically utilized to evaluate circuit timing. Conventional logic synthesis tools leverage STA to further optimize digital circuits and resolve timing issues. They rely on a corner analysis, which means that the circuit timing is evaluated for best and worst process and temperature corners. However, given the extent of variation in the NTC circuits, this approach is too pessimistic. Therefore, Statistical STA (SSTA) should be used to determine the timing of NTC circuits. Conventional SSTA tools propagate random variables, i.e., such as threshold voltage and transistor dimensions, to extract the distribution of output timing. This can be done based on Monte-Carlo analysis, which is extremely time-consuming, or analytical analysis. Based on Equation (4), given that V_{dd} or V_{th} are statistical parameters with a normal (Gaussian) distribution, the resulting delay distribution would be log-normal [43]. Therefore, the SSTA tools used for NTC circuits may need to consider such exponential sensitivity by propagating log-normal distributions, which makes the SSTA tools quite complicated.

3. Related Work

As explained, keeping up with reliability requirements while gaining high energy efficiency is a major challenge for NTC design. Many researchers have focused on these issues to enable widespread use of NTC in various application domains. In this regard, the state-of-the-art can be divided into:

- Design flow and methodology optimization: Due to high integration and the complexity associated with circuit design in modern technologies, Electronic Design Automation (EDA) tools are extensively used to enable circuit design and optimization. Hence, efforts have been focused on improving the EDA tools for NTC. Kaul et al. [33] provided a list of challenges for NTC circuit design, from device modeling to test, and desired design technologies. Recent works have addressed some of the NTC challenges in device modeling [43,72], variation modeling [40,73,74], synthesis [75], leakage power management [75,76], and system-level modeling [77]. There are still major shortcomings in variation-aware library characterization, verification, and testing.
- Design optimization for NTC: Various methods have been proposed to improve the design for NTC, from device-level to architecture-level. At circuit-level, robust standard cell libraries [26], memories [54,56–58] and flip-flops [27,33] are optimized for NTC. Level shifter designs have been vastly studied for interfacing between different voltage islands [60], and clock networks are redesigned for low-voltage operation [73]. Additionally, timing error correction methods such as [49,78,79] can be used to mitigate timing variations at runtime. At architecture-level, caches [80], processor pipeline [81–84] and ISA [85] can be optimized, and other leakage power reduction methods are also investigated [26]. However, there are still many opportunities for cross-layer design optimization.
- Runtime optimization and tuning: Since the MEP is dependent on process and runtime variations, runtime optimization and tuning could be required depending on the running application and the operating environment. Therefore, various runtime tuning methods have been studied, mostly based on a closed-loop hardware-implemented monitoring circuit. It is proposed by many researchers to measure the circuit power online and take actions to maximize the energy efficiency by adapting the supply voltage in a closed-loop feedback [86–90]. Supply voltage and threshold voltage tuning is also recommended to balance the speed of different cores [33,74,91]. However, closed-loop adaptation techniques are associated with additional circuitry for measuring the circuit power and applying the adaptation strategies, which may be too costly for NTC circuits. Therefore, low-cost adaptation methods are highly desirable for NTC.
- Wide-voltage reliability challenges: The impact of aging and supply voltage fluctuation (due to internal activity) is negligible in the NTV region due to low electric field, low current density, and low power consumption; however, operating over a wide-voltage range, from the near-threshold region to the super-threshold region may be required to satisfy performance constraints. In this case, aging phenomena affect the circuit in the super-threshold voltage region, which deteriorates the reliability. Therefore, it is necessary to address such aging challenges when applicable.

Adjusting the supply voltage of the circuit to its MEP depending on the process and runtime variations can improve energy efficiency. However, these techniques do not reduce the idle time of the circuit, when it is not doing any specific task but is wasting leakage power. As the leakage power constitutes a considerable portion of the total power consumption, it is crucial to reduce the idle time of circuits to improve energy efficiency.

Although many methods have been proposed to analytically find the MEP [25,92] or track it during the runtime [86,88,89], finding the MEP point does not necessarily lead to the maximum energy efficiency as a global supply voltage is assigned to the whole circuit (coarse-grained supply voltage assignment). Fine-grained supply voltage assignment methods always lead to better improvement though imposing overheads [75,93].

Performance variations in the NTV region can be addressed by using conservative techniques such as structural duplication, voltage and frequency margining [94]. However, such approaches can impose significant area and energy overhead. Moreover, the conservative timing margin will increase the idle period of the structures and potentially reduce the energy benefits of the NTC. Soft edge clocking [95,96] and body biasing [97] are two other well-known approaches to deal with process variation at NTC. Several methods have already been proposed [75,98], which incorporate synthesis techniques to improve the circuit characteristics for NTC.

Dynamic input vector variation is another source of variation [99]. By applying different input vectors, different parts of the circuits are activated, leading to different propagation delay from inputs to outputs. There are so-called *better-than-worst-case* design techniques to improve circuit performance or energy considering dynamic input vector variations [48–50]. In these approaches, the timing margin of the circuit is reduced to a value smaller than the conservative worst-case margin, and possible timing errors are detected and corrected on-the-fly to achieve error resiliency on top of improved performance and energy. However, traditional designs try to balance the circuit and hence there is a critical point called "path walls" in which reducing the delay beyond this point leads to a huge number of timing failures. In order to reduce timing errors, and hence further improving the efficiency of *better-than-worst-case* designs, re-timing techniques to avoid path walls are introduced in the literature [51,52]. However, such techniques are not effective in NTC because of the increased amount of timing errors.

For some circuits, such as ALU, there is a difference between the execution time of slow instructions (SI) and fast instructions (FI) due to dynamic input vector variation. Based on this fact, an aging-aware instruction scheduling is proposed in [100,101] to execute each instruction group (FI or SI) with its own specialized functional units to improve lifetime.

In NTC, the delay difference between FI and SI is more pronounced. This means that if the clock cycle is set according to SI propagation delay, there is a considerable waste of leakage energy during the execution of FIs. Based on this, we propose a technique in which the clock cycle is significantly reduced close to the propagation delay of FI instructions to reduce the wasted leakage energy. In this approach, SIs are executed in multiple cycles to avoid possible timing failures.

At the super-threshold domain, various coarse-grained power-gating techniques have been proposed to turn-off idle cores [102–104]. The cores are turned-off when determined idle time is observed. However, such approaches require a state preservation mechanism and typically impose a long wake-up latency, which is costly at NTC. Furthermore, two methods based on the power-gating of execution units are proposed in [105]. These techniques allow the power gating of an entire execution unit. However, some of the instructions of an execution unit are utilized rarely by the running applications. Therefore, analyzing the instruction utilization pattern could reveal opportunities for fine-grained power-gating.

4. Cross-Layer Data Path Optimization

4.1. Overview

To save leakage energy, it is important to reduce the time a circuit is powered on but idle, i.e., the time a circuit is performing no operation. For this purpose, the timing slack under all operation conditions has to be trimmed. This is a very challenging task, in particular for functional units, which are fundamental components of data paths. In functional units, different operations have different timing criticalities and slacks. For instance, in an ALU, some operations need only a fraction of a clock cycle (e.g., logic operations), whereas others require a full clock cycle (e.g., addition operation with operands of long data type). Consequently, whenever an operation of the first category executed energy is "wasted" due to leakage, as the clock period is defined according to the slowest operation. Instead, it is much more efficient to execute operations of the second category at multiple cycles, which reduces the overall idle time of the functional unit, as conceptualized in Figure 6.

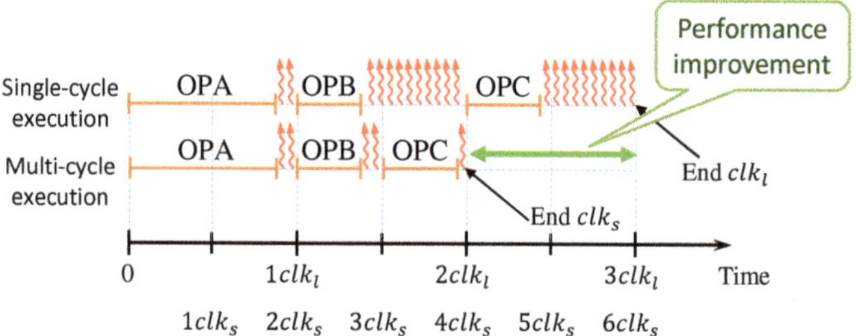

Figure 6. Conceptual illustration of the impact of a short clock period (clk_s) and multi-cycle operations (e.g., OPA) on runtime and "wasted" leakage. Leakage is illustrated by ⸨.

Additionally, when an instruction is being executed by a functional unit, some parts of the functional unit are idle as they are not exercised by the executing instruction. To tackle this issue and improve the overall energy efficiency and reliability, we revisit the structure of the functional unit by partitioning it into multiple smaller and simpler units, to enable fine-grain power gating of unexercised functional units. If a particular functional unit is not utilized by a long sequence of instruction stream, it can safely be power-gated to save leakage energy, as shown in Figure 7. Proper clustering of the instructions into several smaller functional units allows maximizing the power-down intervals of multiple functional units, and hence reducing the leakage energy. At the same time, simplifying the functional unit can reduce timing uncertainties and improve a reliable operation of NTC processors under process and runtime variations.

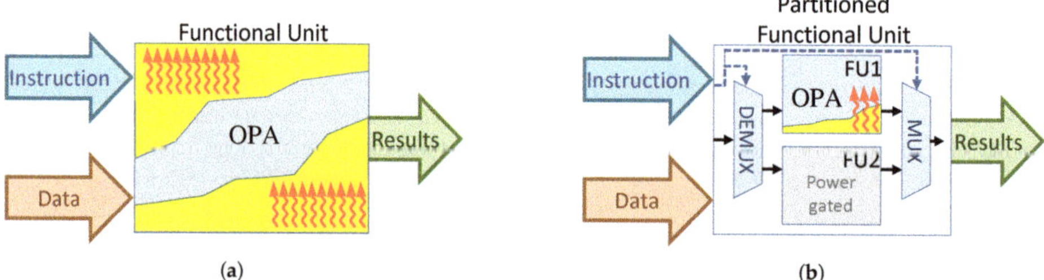

Figure 7. Conceptual illustration of the impact of partitioning on a functional unit executing OPA instruction, and its impact on "wasted" leakage. Leakage is illustrated by ⸨. (**a**) Executing instruction OPA on original functional unit and the associated leakage dissipation by unexercised components. (**b**) Executing instruction OPA on the partitioned functional unit and power-gating of smaller units.

Leakage and dynamic energy are comparable in the NTV region. Therefore, it is crucial to control the amount of leakage power to leverage the benefits of the NTC. This section presents cross-layer methodologies to optimize energy efficiency, performance, and reliability of NTC functional units based on reducing the idle time.

The leakage power can be reduced by reducing the idle time of a circuit. A circuit may become idle within a clock cycle, for example, when it has extra timing slack, or over consecutive clock cycles. The former may be avoided by modifying the circuit design to reduce the timing slack for all conditions, and the latter can be avoided by power-gating the circuit when possible.

The timing-slack minimization can be done by circuit synthesis techniques, which are tailored for NTC circuits [75,98]. However, these approaches are not much efficient when dealing with functional units (e.g., adders, multipliers, or complete ALUs) as the timing slacks of the instructions may be widely different [106]. For example, a simple instruction such as a `Bitwise-AND` only needs a fraction of the clock cycle, whereas an `ADD` instruction could use a large portion of the clock cycle. As the delays of these instructions are intrinsically different, the NTC synthesis techniques are not able to effectively balance the delay of these two instructions. We propose to execute the slow instructions in multiple cycles and the fast instructions in a single clock cycle (instruction multi-cycling) as the solution to this problem [106]. Applying other optimization techniques such as opportunistic circuit synthesis, instruction replacement, and data type manipulation can further improve the effectiveness of the proposed method. The proposed instruction multi-cycling method is described in Section 4.2, the developed methodology is explained in Section 5.1, and its experimental results are discussed in Section 5.4.

In a real-world scenario, an executed application may utilize only a fraction of the instructions implemented inside a functional unit. This means that during the execution of such applications, those gates of the functional unit that are exclusively used by the non-exercised instructions of the functional unit are not utilized. The leakage power from these gates contributes to the dissipated energy of the system. On the other hand, it is not feasible to power off the entire functional unit because any of the instructions can be called at a time. We propose to address this by redesigning the entire functional unit to allow power-gating [107]. In this approach, a large functional unit such as an ALU is partitioned into several smaller functional units such that each unit can be power-gated separately (functional unit partitioning). For this purpose, the instructions need to be clustered properly into different groups. A number of parameters such as the instruction utilization pattern, the temporal distance between the instructions inside an application instruction stream, and intrinsic similarity between the instructions need to be considered for a proper clustering. Accordingly, the instruction stream of various applications has to be analyzed to extract the required information for clustering [107]. The proposed functional unit partitioning method is described in Section 4.3, the developed methodology is explained in Section 5.1, and its experimental results are discussed in Section 5.5.

4.2. Instruction Multi-Cycling

This scheme consists of four main ideas to fully enable the potential of NTC:

1. We classify the instructions into *slow* (with little timing slack) and *fast* (with large timing slack), based on the time required to execute the instructions. The slow instructions are executed in multiple clock cycles, to reduce the leakage power while executing fast instructions within one clock cycle.
2. Contrary to the super-threshold regime, we use relaxed timing constraints for the synthesis of functional units in the NTV region. While this increases the delay of the most critical path, it avoids that all instructions have similar delays and belong to the same category. By that means, the scheme of point 1 is much more efficient.
3. Finally, the clock period is not only set according to points 1 and 2 but also by considering the sensitivity of the functional unit to variations. As a result, it is possible to co-optimize energy, performance, and reliability.
4. In order to reduce the number of "slow" multi-cycle operations, we propose to employ instruction- and compiler-level optimization approaches to replace some of these instructions in the code with fast single-cycle instructions, which further improves energy and performance.

4.2.1. Energy Improvement through Instruction Multi-Cycling

The propagation delay of functional units is typically dependent on input vectors. For example, in an ALU, the instructions can be categorized into slow instruction and fast instructions and the delay gap between fast and slow instructions might be significant. The logical operations such as inversion can be executed very fast while the execution of arithmetic operations such as addition and subtract requires more time. Figure 8a shows the normalized delay of various instructions of an ALU at 0.5 V (according to the simulation setup presented in Section 5). The synthesis tool balances as many paths as possible to optimize the energy given the timing constraints. In this case, 26 instructions are balanced to be critical or near-critical in terms of timing (slow instructions); however, the rest of the instructions require less time for completion (fast instructions).

In a traditional design approach, the delay of the unit is determined by the delay of the slowest instruction. Therefore, during the execution of the fast instructions, the ALU finishes the execution early and leaks for the rest of the clock cycle. The "dissipated" leakage energy is more important in the NTC because (1) the leakage energy is significant and constitutes a significant portion of the total energy, (2) the delay gap between the fast instructions and the slow instructions is even more pronounced due to the impact of process variation. As shown in Figure 8a, the amount of variation induced delay (the difference between the nominal delay and the worst-case delay) is larger for the slow instructions, which significantly increase the delay gap between the slow and fast instructions in the near-threshold regime.

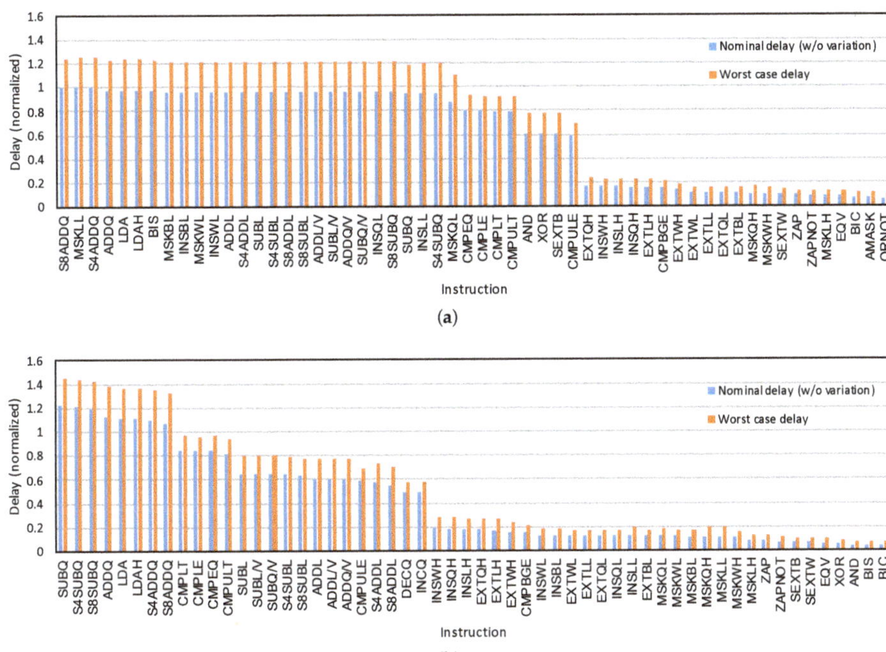

Figure 8. Nominal instruction delay without considering the impact of process variation and worst case instruction delay (considering process variation) are depicted for the instructions of an Arithmetic Logic Unit (ALU) synthesized with (**a**) typical timing constraint (Tight ALU) and (**b**) loose timing constraints (Loose ALU). The worst case delay is extracted by statistical static timing analysis (SSTA) as $\mu + 3\sigma$ of the instruction delay. All the numbers are normalized to the maximum nominal delay of the Tight ALU, which is the nominal delay of S8ADDQ in (**a**). The depicted figures are obtained for a 64-bit ALU based on the simulation setup explained in Section 5.3 at $V_{dd} = 0.5$ V.

Our proposed idea is to reduce the leakage energy by using a shorter clock period and execute the slow instructions in multiple cycles. For example, in the presented ALU in Figure 8a, it is possible to reduce the clock period to half of the delay of the slowest instruction and execute the slow instructions (from S8ADDQ to CMPULE-totally 35 instructions) in two cycles and the rest of the instructions in one cycle. Therefore, once a fast instruction is executed the amount of the wasted leakage energy would be much smaller compared to the traditional approach. This would also improve performance because fast instructions require less time for execution.

For the execution of the slow instructions, which may require more than one cycle, there is no need to insert any flip-flops or latches into the circuit. During the execution of such instructions, the inputs of the circuit are kept unchanged in order to allow the circuit to complete the execution of the slow instructions in more than one clock cycle. Therefore, it is necessary to make some modifications in the microprocessor, as discussed in Section 5.4.5.

We propose a set of cross-layer techniques from logic synthesis to compiler level in order to leverage the maximum benefits from the proposed method by moving more executed instructions to the "fast" category.

4.2.2. Logic Synthesis for NTV Region

In order to show the impact of the logic synthesis on the ALU, we synthesized the same ALU with two different strategies. First, the ALU is synthesized with tight timing and area constraints, which is the default strategy in the super-threshold region. From now on, we refer to this synthesized ALU as *Tight ALU*. As the results depicted in Figure 8a show, the synthesis tool balanced the delay of many instructions (mostly arithmetic instructions) to maximize the energy efficiency and performance. However, there are still many instructions (mostly logic instructions), which are too short to be balanced similar to the slow instructions. Furthermore, there is a sharp transition from the delay of slow instructions to the fast instructions.

Then, the same ALU is synthesized with loose timing and area constraints (*Loose ALU*). The results of this synthesis are shown in Figure 8b. Since the synthesis tool is not under tight constraints, the delays of the slow instructions are larger compared to the Tight ALU, and there is a smooth transition from the delay of the slow instructions to the fast instructions.

In summary, the synthesis results based on the simulation setup presented in Section 5.3, show that the performance of the Tight ALU is better than the Loose ALU by 13% considering the variations; however, due to the wide spectrum of the delays of the instructions in the Loose ALU, there are more opportunities for improving the ALU with the envisioned multi-cycling technique. The reason is that fewer instructions are critical in terms of timing and by cleverly choosing the clock period we can gain in terms of energy and performance, as shown later in Section 5.

4.2.3. High-Level Optimization Techniques

From the above discussion, we observe that some ALU instructions are slower with longer execution time, while other instructions are faster with shorter execution time. Therefore, the aim is to exploit high-level optimization techniques such as data type conversion and instruction replacement to replace slower instructions of an application by fast ones wherever possible, so that the overall energy demand will be reduced further. Since the slower instructions are executed in multiple clock cycles, replacing them with faster instruction can also improve the performance by reducing the Cycle Per Instruction of the applications.

Data type conversion: The selection of data types for variables has a huge impact on the instruction binary generated by the compiler. The data types such as short (1-byte), char (2-byte), int (4-byte) and long (8-byte) not only specify the storage space size of variables but also the type of operation on the variables (e.g., 64-bit arithmetic such as ADDQ vs. 32-bit arithmetic such as ADDL). This affects the occurrence rate of slower and faster instructions within an application. Hence, in addition to memory optimization, a clever data type selection will help to save energy by using fast instructions (e.g., ADDL in Figure 8b) instead of the slow ones (e.g., ADDQ in Figure 8b).

Instruction replacement: Many applications such as sorting and matrix multiplication algorithms spend most of the execution time in loops. Hence, simple instructions such as increasing loop counters and array indexes significantly contribute to the overall instruction count. If such instructions are not assigned appropriate data type and operation, they can impose significant performance and energy overhead. For instance, since modern ALUs have a dedicated increment/decrement circuitry, the usage of increment/decrement instructions plays a vital role in improving the performance and energy efficiency by reducing the number of slower instructions. This and other instruction replacements (e.g., shift instead of multiplication) can be achieved either by the programmer or using different compiler optimization techniques.

4.3. Functional Unit Partitioning

Fine-grained power-gating has been known as an effective solution for reducing the leakage power consumption of a system by cutting the supply lines of the idle components [105]. However, it is costly to power-on and power-gate the components in terms of execution time as each power-gating cycle mandates a minimum time between power-on and power-off cycles. Once a component is put to sleep, its functionality cannot be used. Therefore, the components should be carefully tailored towards power-gating.

One opportunity for power-gating of the components is to determine the unused parts of a circuit based on the running application and power-gate those specific parts instead of the entire component. For ALUs executing several instructions, unused parts of the ALU corresponding to the instructions not executed for a long time could be safely power-gated. Figure 9 presents the usage frequency of a 64-bit ALU (for Alpha ISA). As shown in the figure, there are orders of magnitude differences between the usage frequency of a highly used instruction such as LDA and a rarely used instruction such as ZAP. If both of the mentioned instructions are implemented in a single ALU, the circuitry for ZAP is leaking most of the time while being at an idle state. In other words, the gates that are implemented exclusively for the rarely used instructions are in the idle state most of the time and are contributing to the total leakage of the ALU.

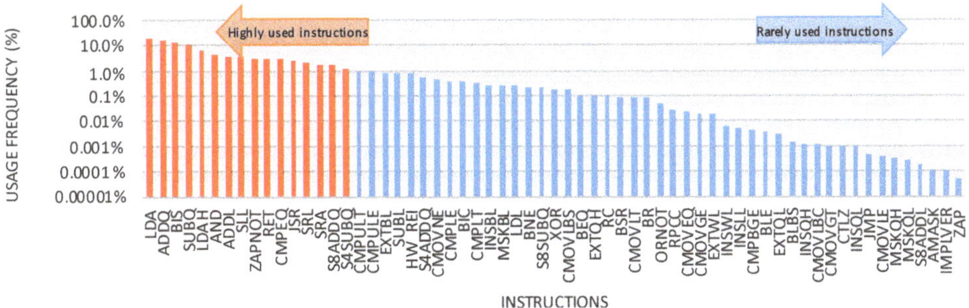

Figure 9. Instruction usage frequency in a 64-bit ALU for gzip workload. There are orders of magnitude differences between utilization frequency of "highly used instructions" on the left and "rarely used instructions" on the right. The depicted figure is obtained by executing SPEC2000 workload "gzip" using the gem5 simulator, as explained in Section 5.3.

Our proposed solution is to partition a functional unit like an ALU into multiple smaller ALUs. This allows us to power-gate some of the ALUs based on the utilization of their instructions by turning off the ALUs that are not currently being used. The original ALU could be partitioned based on different parameters such as the instruction utilization frequency, instruction similarity, and instruction temporal proximity.

In summary, this scheme synergistically exploits NTC in conjunction with a fine-grained power-gating of functional units to enable energy-efficient operation of devices designed for IoT applications. A hierarchical clustering algorithm groups the instructions into smaller functional units by considering the frequent instruction sequences, obtained by application profiling, in order to maximize power-gating intervals. Accordingly,

1. We characterize the instruction flow of representative workloads and analyze the instruction stream in order to obtain instructions' utilization frequency and temporal distance.
2. The instructions are then partitioned into several groups according to the metrics extracted from the instruction stream analysis as well as the inherent similarity of the instruction.

3. Each set of instructions residing in the same partition will be implemented by a dedicated functional unit, and they form a complete functional unit altogether.
4. To reduce the leakage power, only the functional unit corresponding to the running instruction partition is activated while other units are power-gated.

4.3.1. Instruction Pattern Analysis

A careful analysis of the instruction patterns on a set of representative workloads provides information regarding the ALU partitioning. For this purpose, we first simulate the execution of a set of representative workloads by an architectural simulation tool as explained in Section 5, and then based on the extracted instruction streams the utilization frequency and the temporal distance of different instructions are extracted.

Instruction utilization frequency: The instruction utilization frequencies presented in Figure 9 show a significant difference in utilization among the instructions. The instruction utilization frequency has an inverse relation with the power-gating feasibility for the ALUs associated with the instructions. For example, the ALUs that contain instructions ADDQ and BIS are less likely to be power-gated because these instructions appear in the instruction buffer on average every 10 cycles. However, it is more likely to power-gate the ALUs implementing the rarely used instructions on the right side of Figure 9 such as S8ADDL.

The utilization frequency of an instruction can be simply defined as the number of cycles in which the instruction is executed divided by total cycles. For a given instruction stream $\mathbf{S}$, which is a sequence containing $N = |\mathbf{S}|$ instructions, we can define the frequency of instruction A as:

$$Freq_A = \frac{\#S_i \in \mathbf{S} \ such \ that \ S_i = A}{N} \tag{5}$$

where S_i is the i-th element (instruction) of $\mathbf{S}$. If an instruction is used rarely, it can be easily grouped into any existing ALUs. However, we need to be more cautious in grouping frequently used instruction because the grouping strategy might improve or deteriorate the power-gating capability of the ALUs. Based on this analysis, we can define the frequency distance metric between two instructions A and B as the geometric mean:

$$dist_{A-B}^{freq} = \sqrt{Freq_A * Freq_B}. \tag{6}$$

Such definition facilitates the partitioning of rarely used instructions into a single ALU.

Instruction temporal distance: Some instructions are more likely to appear next to each other in an application. For example, from "bzip2" workload, we observed that ADDL appears after LDA on average every 2.97 instructions (see Figure 10). In these cases, it could be beneficial to group these neighboring instructions inside one ALU in order to improve the power-gating interval for other ALUs.

The temporal distance between two instructions A and B can be extracted based on the number of cycles between any occurrences of A and B. For example, according to the results presented in Figure 10, there are 19146 LDA instructions that are directly followed by an ADDL instruction, and there are 11860 LDA instructions that are followed by an ADDL instruction after three cycles. Based on the workload analysis, a distribution is extracted for every instruction pair A–B, which explains the percentage of the A–B pairs that are far from each other by k cycles. The Survival Function (SF) (defined as $1 - CDF$ of a distribution) of these *temporal distance* distributions is useful in the clustering problem definition in Section 4.3.2.

We define a *temporal distance set* which contains the indexes of consequent instructions A and B and their corresponding distances (i, k):

$$Temporal_{A-B} = \left\{ (i,k) \mid 0 < i, \ 0 < k, S_i = A, \ S_{i+k} = B, \nexists j, \ 0 < j < k, \ S_{i+j} \in \{A, B\} \right\}. \tag{7}$$

Accordingly, the Probability Mass Function (*PMF*) based on distance k is calculated as:

$$PMF_{\text{A-B}}(k) = \frac{\left|\{(i,r)|(i,r) \in Temporal_{\text{A-B}}, \; r = k\}\right|}{|Temporal_{\text{A-B}}|}. \tag{8}$$

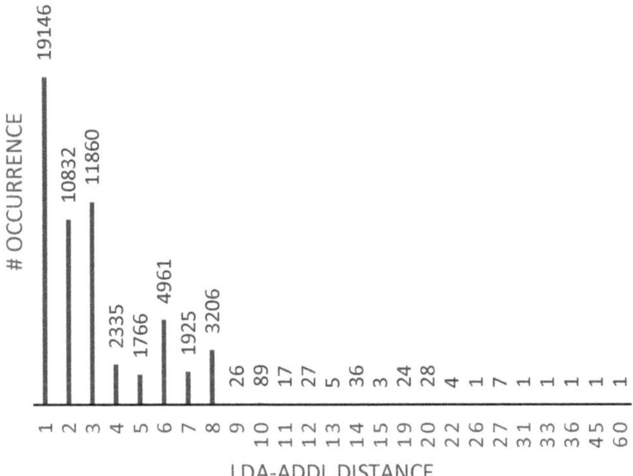

Figure 10. The temporal distance between LDA and ADDL instructions in "bzip2" workload (simulation for 2 million cycles). There are 19,146 cases in which the ADDL instruction appeared right after LDA. The average distance is 2.97. The results are obtained using gem5 simulator as explained in Section 5.3.

The extracted *PMF* is then used to find the *SF* as follows:

$$CDF_{\text{A-B}}(k) = \sum_{i=1}^{k} PMF_{\text{A-B}}(i), \tag{9}$$

$$SF_{\text{A-B}}(k) = 1 - CDF_{\text{A-B}}(k). \tag{10}$$

In a fictitious scenario where only instructions A and B exist, and they are divided into two ALUs, the *SF* can explain the power-gating possibility. In this case, if the minimum number of cycles required to perform a power-gating (power-gating threshold) is $PGTH$, then $SF_{\text{A-B}}(PGTH)$ obtains the power-gating probability. Therefore,

$$dist_{\text{A-B}}^{temporal} = SF_{\text{A-B}}(PGTH) \tag{11}$$

can be used as the temporal distance metric between two instructions A and B.

Instruction similarity: Many instructions share some gates in ALU mostly due to their similarity. For example, an ALU could have different addition and subtraction instructions, which are inherently similar. As a result, these instructions share a large portion of gates in the synthesized netlist. Therefore, implementing these instructions in separate ALUs would impose redundant structures leading to undesirable leakage and area overhead. Therefore, it is preferable to group such instructions into one ALU to reduce the associated overheads.

We introduce a dissimilarity metric defined as the structural dissimilarity between instructions ($dist_{\text{A-B}}^{dissimilarity}$). For example, $dist_{\text{ADDL-ADDQ}}^{dissimilarity} = 0.0$ as both addition instructions implement similar functionality. However, $dist^{dissimilarity}_{\text{ADDL-ORNOT}} = 1.0$ as the corresponding instructions implement two completely different logic structures. The dissimilarity values are assigned based on the knowledge we have about the logic implementation of different instructions.

Some of the above parameters may lead to contradictory grouping of instructions into ALUs. For example, instructions S8ADDL and ADDL should be grouped into one ALU because of inherent similarity; however, according to their utilization frequencies they should be placed into different ALUs to allow power-gating. In the next section, we define a formal clustering problem considering the aforementioned parameters and solve it to find the best instruction grouping strategy.

4.3.2. Instruction Clustering Problem Definition

The problem of partitioning a large ALU into smaller ALUs can be defined as a clustering problem, in which the distance between the instructions is explained by temporal proximity, utilization frequency, and similarity of instructions. The goal of such clustering algorithm is to maximize the distance between ALUs while minimizing the distance between instructions of each ALU. This allows us to increase the overall power-gating likelihood of ALUs, which leads to lower leakage and better energy efficiency.

For this purpose we apply the Agglomerative Hierarchical Clustering (AHC) algorithm [108] to cluster the instructions into several groups, each group implemented in one ALU. AHC is suitable for our problem because we can provide pairwise distances between each and every two instructions.

We create the pairwise distance matrix needed for the AHC algorithm based on the frequency distance metric ($dist^{freq}_{\text{A-B}}$), temporal distance metric ($dist^{temporal}_{\text{A-B}}$), and structural similarity ($dist^{similarity}_{\text{A-B}}$) introduced in the previous section. Finally, the elements of the pairwise distance matrix ($pdist$) are obtained as (Cartesian distance on a 3D space):

$$pdist^2_{\text{A-B}} = (\beta.dist^{freq}_{\text{A-B}})^2 + (\gamma.dist^{temporal}_{\text{A-B}})^2 + (\lambda.dist^{dissimilarity}_{\text{A-B}})^2. \tag{12}$$

Here, β, γ, λ are coefficients to scale all the metrics into the same scale.

We consider the single-linkage clustering method on the AHC. In a single-linkage method, the linkage function $D(X, Y)$, which is the distance between two clusters X and Y, is defined as the minimum distance between every two members of the clusters:

$$D(X, Y) = \min_{A \in X, B \in Y} pdist_{\text{A-B}}. \tag{13}$$

Therefore, maximizing the distance between clusters X and Y will allow the maximum power gating possibility of the corresponding ALU implementations and improves the energy efficiency.

4.3.3. Fine-Grained Power-Gating Prediction

Once the inactive phase for a component is detected at architecture-level, the component can be power-gated by asserting a sleep signal on the header/footer sleep transistors. Although the power-gating can effectively reduce the wasted leakage energy, it has to be done when the functional unit is not utilized for a minimum number of cycles ($PGTH$) to break-even the associated overheads. This value is estimated to be around 10 cycles for a typical technology [105,109].

The inactive phase of a functional unit can be predicted based on several techniques at runtime. In order to determine the inactive interval of each functional unit and decide whether to power gate it or not, it is possible to monitor the instruction buffer for a number of upcoming instructions while considering the branch prediction buffer. However, the power-gating signal for a given functional unit can be mispredicted due to branch misprediction. As a result of misprediction, the entire pipeline may be needed to be flushed to reload the correct instructions. This provides some time to properly power up the required functional units without imposing much overhead due to pipeline stall. It is worth mentioning that sub-threshold and near-threshold processors typically have a deeper pipeline to benefit in terms of performance and energy efficiency [110]. In such processors, there is enough time for functional unit power up after misprediction due to deeper pipeline design.

In Out of Order (OoO) processors, the order of execution of the instructions can slightly change at runtime in order to avoid stalls in the pipeline. As a result, predicting the idle time of the functional unit partitions is not straightforward. The proposed functional unit partitioning method can be used whenever such prediction is possible; however, without good prediction methods, the proposed method may not bring significant improvement to the design. In the presented experimental results in Section 5.5, we evaluated the results with the assumption of in-order execution of the instructions. Further development and application of the proposed functional unit partitioning method to OoO designs is not presented in this paper.

5. Results and Discussion

This section presents the methodology and simulation setup for the proposed data path optimization approaches. We evaluate the effectiveness of the proposed approaches by applying them to an ALU.

5.1. Implementation Flow

5.1.1. Input Vector Dependent Timing and Power Analysis

As explained in Section 4.2, the time and power required by a functional unit to complete an operation execution is different from one instruction to another. Here, we use the flow illustrated in Figure 11a to extract the detailed timing and power information for each instruction. Therefore, for an ALU as a case study the following steps are performed:

1. Synthesis: The synthesis step of Figure 11a is executed with timing constraints to obtain the gate level netlist of the synthesized ALU.
2. Netlist modification: The synthesized netlists of the ALUs are modified such that the OPCODE input signals that determine the instruction to be executed are changed to internal wires. For each instruction, the OPCODE signals are assigned to associated values inside the Verilog netlist. This will effectively deactivate the rest of the instruction paths in the ALU and force the STA tool to evaluate only the paths associated with the execution of the given instruction. This is because the rest of the paths (which belong to other instructions) are deactivated, and any change in the output pins of the ALU is only due to the paths belonging to the given instruction.
3. Timing and power analysis: The timing and power analyses are performed on the modified netlists, and the delay of the instructions considering the variation as well as the dynamic power and the leakage power for each instruction are extracted.

The change in the leakage power from one instruction to another is negligible because even the inactive gates that are not part of the propagation paths are leaking. However, if the execution time is different from one instruction to another, the amount of the leakage energy would be different proportionally. The dynamic energy for each instruction is also calculated based on the dynamic power value.

5.1.2. Functional Unit Partitioning Flow

Figure 11b shows the overall flow of the proposed functional unit partitioning method applied to an ALU. The flow consists of three distinct steps:

1. Instruction Pattern Analysis: In this step, the instruction stream extracted from running representative workloads are analyzed to extract instruction temporal distance and utilization frequency. Instruction dissimilarity is also defined based on the field knowledge about the implementation of logic units.
2. Instruction Clustering and ALU generation: With the help of the information collected from the previous step, instructions are grouped into *n* clusters, and each cluster is implemented as a new ALU. Moreover, all the partitioned ALUs are combined with a circuit for multiplexing their outputs to form a *union ALU*, as shown in Figure 7b.
3. Energy, Performance and Reliability Analysis: The union ALU generated from the previous step is evaluated in terms of timing and performance, and finally, its reliability is evaluated when it is used instead of the original ALU.

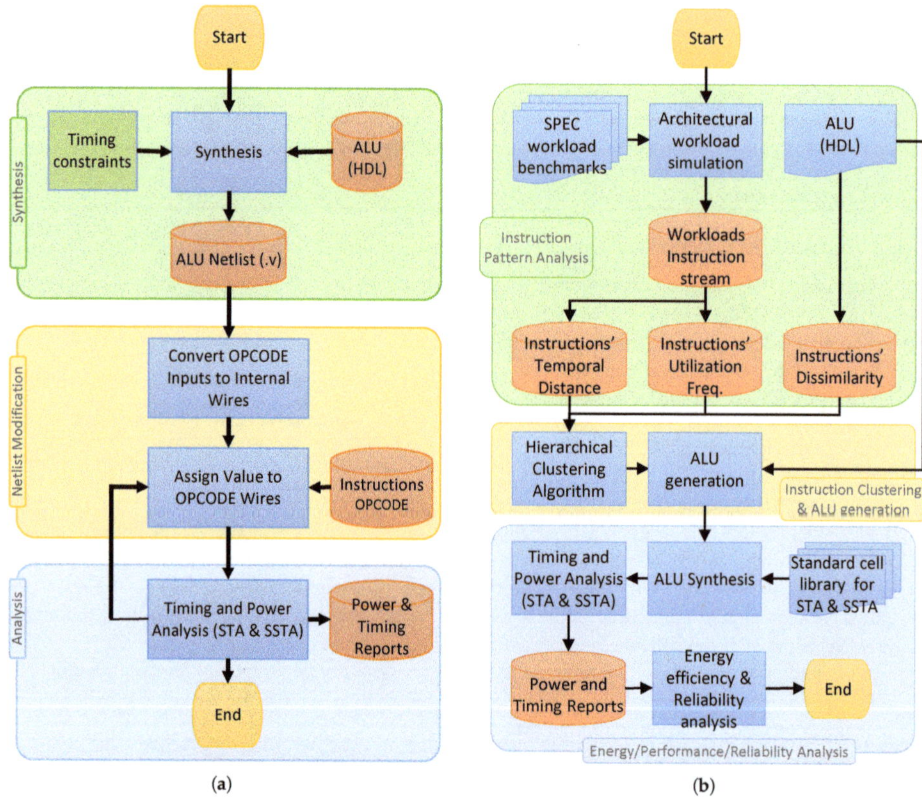

Figure 11. Implementation flows of (**a**) instruction multi-cycling, (**b**) functional unit partitioning applied to an ALU. (**a**) Implemented flow for obtaining the timing and power information of each ALU instruction, in the instruction multi-cycling approach. (**b**) Implemented flow of the proposed functional unit partitioning for ALU optimization.

5.2. Reliability Analysis

Various sources of delay variation, for example, process variation, aging, temperature and voltage variations, can potentially lead to timing failures. Therefore, timing failure is presented as a stochastic metric, which is dependent on the value of an additional timing margin and hence the allocated clock period. Therefore, statistical information regarding the circuit delay can be used to evaluate the reliability. Accordingly, *Reliability* is the probability of not having a timing failure due to variation effects.

In a functional unit such as an ALU, the Failure Probability of a circuit caused by timing issues can be modeled as a function of the allowed time for instruction execution T, based on the delay distributions of the instructions:

$$Reliability = CDF_{ALU}(T) = \prod_{\text{INST}}^{\{\text{all instructions}\}} CDF_{delay,\text{INST}}(T),\tag{14}$$

where $CDF_{delay,\text{INST}}$ is the Cumulative Distribution Function (CDF) of delay of instruction INST. Accordingly, the *Failure Probability* is obtained as:

$$Failure\ Probability = 1 - Reliability.\tag{15}$$

There is a trade-off between reliability and performance as explained in the above equation. A larger clock period (T) results in higher reliability and lower failure probability at the cost of speed.

In a multi-cycling scenario, the allowed time for instruction execution T is dependent on the number of cycles allocated by each instruction. For a single-cycle instruction T is equal to the clock period T_{clk}; however, a two-cycle instruction is allowed to be executed for $2T_{clk}$. Therefore, Equation (14) is modified as follows:

$$CDF_{ALU}(T_{clk}) = \prod_{\text{INST}}^{\{\text{all instructions}\}} CDF_{delay,\text{INST}}(n_{\text{INST}} \times T_{clk}).\tag{16}$$

In the above equation, n_{INST} is the number of cycles allocated for instruction INST obtained based on the distribution of the instruction delays:

$$n_{\text{INST}} = \lceil \frac{d_{\text{INST}}}{T_{clk}} \rceil.\tag{17}$$

d_{INST} is the instruction delay considering the variation, i.e., a point in the tail of the delay distribution referring to very low failure probability. Please note that slow instructions have large logic depth, i.e., there are many gates in the critical paths of these instructions. According to the *Central Limit Theorem* [111], the delay distribution of such instructions is approximately normal (*Gaussian*). In such case, we can use parameters such as the mean (μ) and standard deviation (σ) of instruction delay to approximate its CDF function. Therefore, we may choose $\mu + 3\sigma$ of the instruction delay as d_{INST}, which corresponds to less than 0.135% failure probability.

We perform SSTA to evaluate the impact of process variation [112] on the timing of the circuit. The SSTA tool reads variation information from the variation library (see [71]) containing variation information at the cell-level. The SSTA extracts accurate delay distribution for each instruction of the ALU represented by their CDF (i.e., $CDF_{delay,\text{INST}}$). Based on these distribution functions and Equation (16), we extract the reliability of the ALU.

5.3. Simulation Setup

The proposed approaches are applied to a 64-bit ALU, namely the ALU of the Illinois Verilog Model (IVM) [113], which is a Verilog model for the Alpha 21264 core. The selected ALU implements 64-bit RISC Instruction Set Architecture of Alpha processor. For this purpose, we synthesize the HDL from the IVM ALU with loose and tight timing constraints and characterize it for the NTC by an SSTA to extract the power and delay of each instruction as well as the reliability as explained in Section 5.1. The synthesis is done using the Synopsys Design Compiler [114]. Statistical static timing analysis (SSTA) is done using a Cadence Encounter Timing System [115], which is able to perform SSTA using statistical CCS or ECSM standard cell library models [116].

In order to prepare such statistical standard cell libraries for timing and power analysis, the cells from selected libraries [117,118] from 32 nm down to 10 nm are re-characterized using the Cadence Virtuoso Variety [119] and the cells that are not suitable for NTC are excluded from the library [33]. The Variety tool characterizes the cells considering the process variation and all the power and delay values are stored in the generated standard cell libraries. The process of library characterization can be time consuming. However, this is only done once and the obtained libraries can be used in different projects.

The SSTA performed by this method is applicable to the NTV region, because it can consider a skewed delay model for the cells. Each SSTA timing analysis run takes more execution time compared to a simpler STA run, which is typically done at nominal supply voltage ranges. Nevertheless, it is still much faster compared to a Monte-Carlo-based SSTA. The same method is applied to synthesize and analyze the partitioned ALUs generated by the functional unit partitioning method.

Additionally, we extract the instruction stream for SPEC2000 benchmark workloads using a gem5 architectural simulator [120]. This is done by executing the workloads using gem5, fast forwarding to the start of the executed workloads, and collecting all the executed instructions and the provided operands. The result of the architectural simulation is used to evaluate the impact of the proposed instruction multi-cycling approach and to perform instruction pattern analysis and clustering in the proposed functional unit partitioning approach. The results of both approaches are compared to the baseline, which is an IVM ALU that is synthesized with conventional tight timing constraints.

5.4. ALU Multi-Cycling Results

5.4.1. Multi-Cycling Improvement

By reducing the clock period below the delay of the slowest instruction, some instructions need to be executed in multiple clock cycles. The number of the required clock cycles can be calculated from Equation (17). The delay of the slowest instruction is 146 ns; hence, the minimum clock period for running all instructions in one cycle is also 146 ns.

We changed the clock period of the Loose ALU and extracted the instructions, which should be executed in multiple cycles for each clock period. Here, the clock period is swept from 49 ns (one-third of the delay of the most critical instruction) to 100 ns to explain the impact of clock period on the circuit characteristics. As shown in Figure 12a, some instructions need to be executed in three cycles when the clock period is below 73 ns (which is half the delay of the most critical instruction—marked by a vertical dashed line in the figure).

The energy and performance improvement of the ALU executing workload "equake" are plotted in Figure 12b for various clock periods. The improvement numbers are calculated in comparison with the Tight ALU, which executes all instructions in one clock cycle. When the clock period is reduced, the energy and performance improvements increase because the delay slacks of the instructions are trimmed. However, when any change in the sets of 1-cycle, 2-cycle or 3-cycle instructions happens due to the clock period reduction, i.e., some 1-cycle instructions become 2-cycle instructions or some 2-cycle instructions need to be executed in three cycles, the improvements suddenly drop. The maximum energy improvement (34%) and performance improvement (19%) are achieved when the clock period is 49 ns.

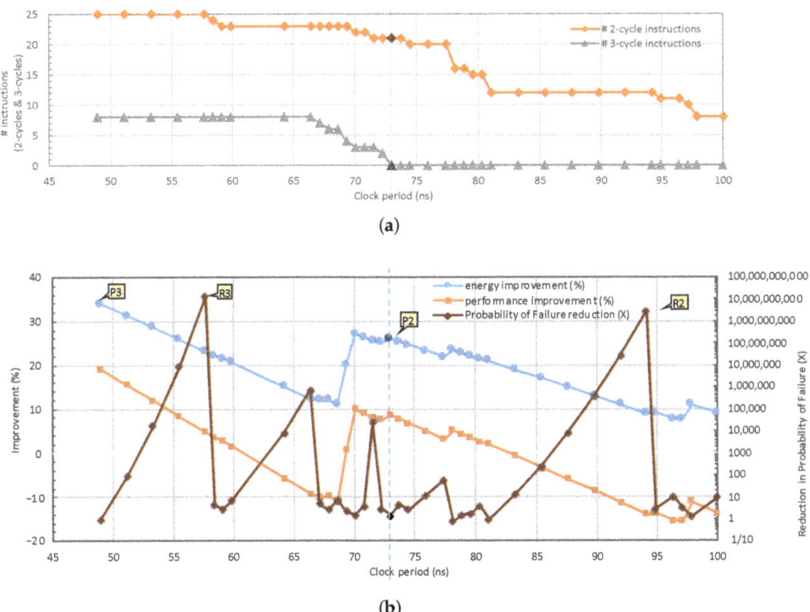

Figure 12. Modifying the clock period of the ALU changes the set of 2-cycle and 3-cycle instructions, which impacts the energy, performance and reliability improvements significantly. The results are extracted for Loose ALU + INC/DEC running workload "equake" at $V_{dd} = 0.5$ V (for other workloads the improvement plots show a similar trend). The improvement values are relative to the Tight ALU. Four Pareto points are marked on the graph: P2 (R2) provides the best energy and performance (best reliability) when ALU is limited to execute all instructions in two clock cycles, and P3 (R3) provides the best energy and performance (best reliability) when some instructions can be executed in three clock cycles. (**a**) Number of 2-cycle and 3-cycle instructions for different clock period values. (**b**) Energy, performance and reliability improvements for different clock period values.

5.4.2. Impact of Workload on the Improvement Ratio

Executing a slow instruction has no performance or energy benefits because it utilizes the ALU for several clock cycles. However, executing a fast instruction, which can be executed in fewer clock cycle(s) will save some energy and improve performance. Since each workload executes a specific set of instructions, the amount of improvement is highly dependent on the profile of the instructions utilized by a workload. Workloads that frequently execute fast instructions will have better improvements using the proposed approach. We measured the amount of energy improvement for different workloads, as shown in Figure 13. The clock period is once set to 73 ns, which means all instructions are executed in at most two clock cycles, and then set to 49 ns to evaluate the results for when the instructions can occupy three clock cycles. The hatched bars in this figure show the amount of energy improvement in different workloads compared to the baseline. The improvement is smaller for workloads that frequently execute slow instructions (applu, mgrid, swim), and larger for workloads with more executions of fast instructions (equake, mesa, mcf, twolf). For example, 76% of the executed instructions in "applu" are among the eight slowest instructions (`addq`, `lda`, `ldah`, `s4addq`, `s4subq`, `s8addq`, `s8subq`, `subq`, as depicted in Figure 8b). However, this is only 45% for workload "equake". The average energy improvement and the performance improvement over all workloads are 20.8% and 1.7%, respectively.

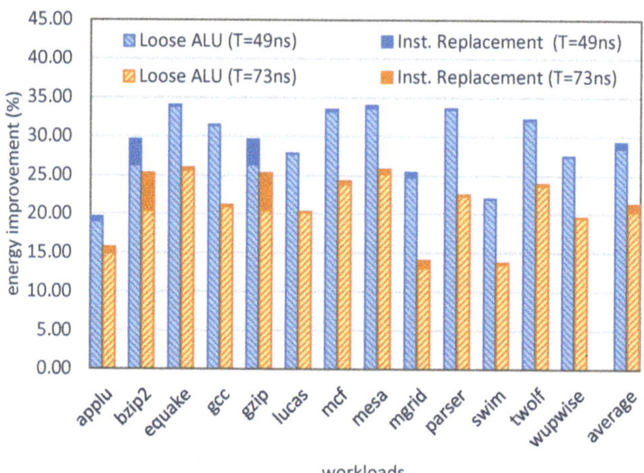

Figure 13. Energy improvement over the baseline (Tight ALU). The additional improvement is also calculated for when addition/subtract instructions are replaced by increment/decrement instructions when possible (clock period is 73 and 49 ns).

5.4.3. High-Level Optimization Improvements

Whenever fast instructions are more utilized by a workload, the energy improvements are even higher. This concept can be incorporated at a higher level to perform application and compiler optimization, which further improves energy efficiency and performance. This can be done in various ways, as discussed in Section 4.2.3.

Our analysis of the ALU instruction stream shows that in some workloads, a number of `add`/`subtract` instructions can be replaced by `increment`/`decrement`. In workloads such as "bzip2" and "gzip", which utilizes `add` and `subtract` a lot, it is possible to change up to 11% of all instructions to `increment`/`decrement`. Since `increment`/`decrement` instructions are faster compared to `add`/`subtract` instructions, the improvement in energy and performance is considerable for these types of workloads. The energy improvements and the boost from applying the instruction replacement are depicted in Figure 13 for two clock periods: 73 ns (all instructions are executed in at most two clock cycles) and 49 ns (three clock cycles). For 49 ns, the energy and performance improvements are 3% and 4.3% higher for "bzip2" and "gzip" when the instruction replacement technique is applied. The technique is still applicable to other workloads; however, the improvements are less (approximately 1%). Applying the instruction replacement technique increases the average energy and performance improvements to 29.3% and 12.8% when the clock period is 49 ns (21.4% and 2.4% on average when the clock period is 73 ns).

In order to show the benefits of data type conversion, we executed a simple "matrix manipulation" application, which calculates $M1 + M2 * 2$ for given matrices $M1$ and $M2$. The corresponding results are presented in Table 1 for two clock periods: 73 and 49 ns. With a clock period of 49 ns and 2-byte data type used for "matrix manipulation", the energy and performance improvements of the multi-cycled Loose

ALU over the baseline are 26.5% and 8.6%, respectively. However, changing the data type to a 1-byte data type increases the energy and performance improvements to 29.0% and 12% over the baseline.

Table 1. Energy and performance improvements of executing the "matrix manipulation" workload with different data types.

Clock Period	Data Type	Baseline		Proposed Multi-Cycling Approach			
		Energy (nJ)	Time (μs)	Energy (nJ)	Time (μs)	Energy Improvement	Performance Improvement
73 ns	short (2-bytes)	27.9	502	22.2	502	20.3%	0.1%
	char (1-byte)	22.5	404	17.6	396	21.5%	2%
	Overall improvement (baseline-short → multi-cycling-char)					36.9%	21.1%
49 ns	short (2-bytes)	27.9	502	20.5	459	26.5%	8.6%
	char (1-byte)	22.5	404	15.9	356	29.0%	12%
	Overall improvement (baseline-short → multi-cycling-char)					42.9%	29.2%

Furthermore, the 1-byte data type is inherently more energy-efficient compared to the 2-byte data type (22.5 nJ vs. 27.9 nJ). Therefore, the cumulative energy improvement of changing the data type from 2-byte data type to 1-byte data type is 42.9% (going from 27.9 to 15.9 nJ). Additionally, the performance also improves by 29.2% (going from 502 to 356 μs).

In summary, high-level optimization methods, such as instruction replacement and data type conversion, can be used to increase the benefits from the proposed instruction multi-cycling. However, the energy and performance improvement obtained by these methods is highly dependent on the executed workload.

5.4.4. Energy/Performance/Reliability Trade-Off

In the near-threshold voltage region, each instruction has a much wider delay distribution compared to the super-threshold region with a longer tail. Therefore, changing the clock period affects the tail of the distribution contributing to failures. We calculate the failure probability for each clock period according to Equation (15) and obtain the *reliability improvement* as the ratio of the failure probability between the baseline and the optimized ALU:

$$\text{Reliability improvement } (T_{clk}) = \frac{\text{Failure Probability of baseline ALU } (T_{clk})}{\text{Failure Probability of optimized ALU } (T_{clk})}. \tag{18}$$

The reliability improvement is depicted in Figure 12b versus the clock period. There are points where the reliability improvement worsens, which are mostly points with very high energy and performance improvement. This is due to the fact that for these points of clock periods the timing margins of most of the instructions are very small, leading to a significant probability of failure even larger than the baseline. However, there are several points with orders of magnitude better reliability. For example, the reliability improvement ratio is $\approx 1.3 \times 10^{10}$ when the clock period is 57.7 ns or 22,891 when the clock period is 71.5 ns. The performance and energy improvement is also significant in these points. The reason for such large reliability improvements is that the baseline has many critical or near-critical instructions with zero or minimal slack. However, the multi-cycling strategy is able to provide enough timing margin for many instructions such that the provided time for the execution of each instruction marks very high sigma values on the tails of all the delay distributions. Therefore, the designer is able to find a good trade-off among energy efficiency, performance and reliability according to the design requirements.

5.4.5. Discussion

It is evident from the results analyzed before that the multi-cycling approach can provide considerable energy and performance benefits, as well as reliability improvements. However, these results represent the ideal case where the clock period can be adjusted without any constraints.

In a real processor (or an ASIC), there may be other timing critical components, limiting the freedom for modifying the clock period. Hence, another operation point in Figure 12b might be chosen. In such scenarios, the entire processor has to operate at a higher frequency in order to exploit a short clock period for the ALU. This may require additional design tweaks for other parts of the processor. For this purpose, timing critical components of the processor can be split up into multiple "short-cycle" components. Consequently, this results in a deeper pipeline, requiring more control logic and more registers. However, this overhead should be compensated by the overall leakage savings of the entire processor due to the runtime benefits of our novel multi-cycling approach. In case a multi-cycle instruction has to be executed by the ALU, the pipeline frontend of the microprocessor is stopped (through using a no-operation instruction, or clock gating), such that the ALU can work for multiple cycles without change of the input vector.

An alternative solution is to execute multiple short instructions in the ALU per clock cycle. By that means, no major modifications of other processor components are necessary. For instance, to execute two instructions in a clock cycle, the high and low levels of the clock signal can be exploited as it was done in the Intel Pentium 4 [121]. Executing more than two instructions in one cycle requires additional shifted clock signals. Therefore, this scheme is only feasible for processors featuring reservation stations to execute multiple instructions per clock cycle.

In an energy efficient design, the operands of a functional unit are loaded from the memory before the functional unit starts executing the instruction, in order to improve power consumption and performance. These operands are typically stored in the local registers, which are easily accessible by the functional unit without much latency. As an example, the registers at the I/O of an ALU in a pipelined processor store the operands used by the ALU.

5.5. Functional Unit Partitioning Results

This section presents the results of applying the proposed functional unit partitioning to an ALU, based on the simulation setup presented in Section 5.3.

5.5.1. Clustering Results

The distance metrics are extracted based on the analysis of the SPEC benchmark workloads. The temporal distance metric is extracted for $PGTH = 100$. The dendrogram in Figure 14 illustrates the results of the AHC method. As shown in this figure, most of the rarely used instructions in Figure 9 are either grouped together or grouped with other closely similar structures. The original ALU can be partitioned into n smaller ALUs according to the dendrogram. One can determine the best n value, based on *inconsistency coefficients* [122]. However, as the overhead of hardware implementation increases with the number of clusters, we only have to limit n to have at most four clusters (ALUs). Here, we present results for partitioning the ALU into three and four smaller ALUs as the energy improvement for two ALUs vanishes away as the power-gating threshold ($PGTH$) increases.

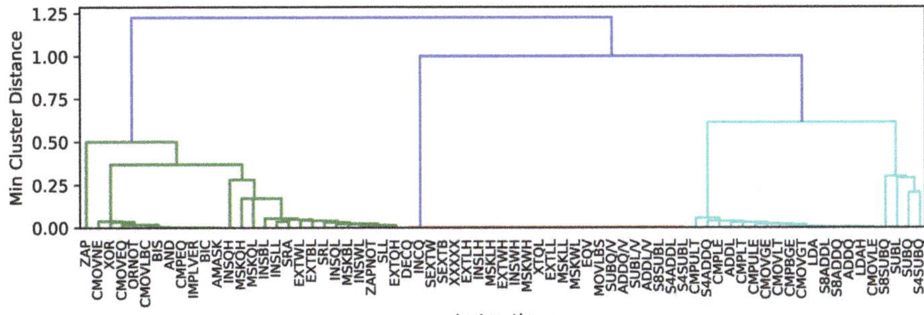

Figure 14. Dendrogram illustration of the proposed AHC method for some of the instructions. The instructions are merged bottom-up to form larger clusters.

5.5.2. Circuit-Level Results

Based on the clustering method, the original ALU is partitioned into three ALUs (3-ALU) and four ALUs (4-ALU). Table 2 reports the area overhead and performance improvement of the 3-ALU and 4-ALU compared to the original ALU at 14 nm. In the performance improvement calculation, we also considered the delay of the extra multiplexer/demultiplexer circuit for both 4-ALU and 3-ALU designs.

Both 3-ALU and 4-ALU designs occupy more area compared to the original ALU as expected. The reported area overhead is the increase in the overall ALU area in percentage, considering all additional circuits. Please note that in modern processors, the die area is dominated by memory elements [123]; hence, the increased ALU area does not contribute to a significant change in the overall die area. On the other hand, these ALUs are faster than the original ALU because the logic implemented in each of the partitioned ALUs is simpler and more coherent. By reducing the supply voltage, the performance gain diminishes slowly, because the impact of process variation on the shorter critical paths of smaller ALUs is more than the long critical path in the original ALU.

Table 2. Energy improvement results for 3-ALU and 4-ALU over Original ALU for 14 nm PTM [117].

	Area Overhead (%)	Performance Improvement (%)	$V_{dd}(V)$	Energy Saving for Power-Gating Threshold ($PGTH$) in %				
				10 Cycles	20 Cycles	50 Cycles	100 Cycles	500 Cycles
3-ALU	17	11	0.80 (super-V_{th})	22.8	19.7	19.5	19.4	17.7
		7.0	0.35 (near-V_{th} *)	24.6	21.4	21.2	21.1	19.2
		5.5	0.25 (sub-V_{th})	27.7	24.4	24.1	24.0	21.8
4-ALU	19	11	0.80 (super-V_{th})	43.1	38.6	32.1	27.6	15.1
		7.0	0.35 (near-V_{th})	43.4	38.9	33.0	28.9	17.3
		5.5	0.25 (sub-V_{th})	44.1	39.6	34.5	31.1	20.6

* In this technology V_{th} is close to 0.35 V.

Additionally, the overall energy improvement achieved by the proposed ALU partitioning method is extracted and reported in Table 2. This is done by calculating the percentage of the time that each smaller ALU can be power-gated based on a given power-gating threshold ($PGTH$) and for each workload. As presented in Table 2, there is a significant energy improvement even in the super-threshold region

(V_{dd} = 0.80 V). The reason is that one of the partitioned ALUs is mostly in sleep mode because its instructions are rarely used (see Figures 9 and 14). The percentage of the time an ALU is power-gated for a specific $PGTH$ is the same for all supply voltages; however, the energy improvement by the proposed ALU partitioning technique is slightly more at lower supply voltages. The reason is that the leakage power contribution to the overall power consumption grows significantly by reducing the supply voltage. Therefore, reducing the same amount of leakage results in a larger percentage of energy saving at lower supply voltages.

Figure 15 compares the energy improvement results for 4-ALU at near-threshold region (0.35 V) in two technology nodes: 10 and 14 nm. As shown, the energy improvement results for these technology nodes resemble each other closely. A similar trend is observed for the rest of the results.

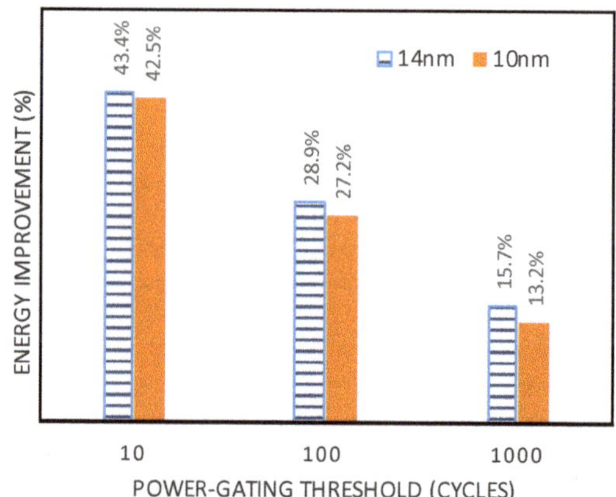

Figure 15. Energy improvement of functional unit partitioning, on an ALU partitioned into 4 smaller units (4-ALU), for 10 and 14 nm technology nodes.

5.5.3. Performance and Reliability Trade-Off

The performance improvement of the functional unit partitioning, shown in Table 2, could be traded for reliability at low supply voltages, similar to what was explained for instruction multi-cycling in Section 5.4.

According to the results, it is possible to enjoy maximum performance improvement (for example 11% at 0.80 V) or invest the achieved speed on reliability to get up to 10^9 times lower failure probability. At the near-threshold supply voltage (0.35 V) the maximum achievable reliability improvement is 11.5× (by trading 7% performance improvement from Table 2) because the delay distribution of the union ALU becomes wider than the original ALU due to its shorter critical path.

6. Summary

The assumptions and optimization targets for NTC circuit design are different from the conventional super-threshold design, due to large impact of variabilities as well as comparable contributions of leakage power and dynamic power. This paper presented two cross-layer design optimization approaches for NTC circuits to co-optimize energy-efficiency, reliability, and performance.

In the *instruction multi-cycling* approach, the idle time of the functional units is reduced by executing slow instructions in multiple clock cycles and fast instructions in one clock cycle. This approach consists of circuit redesign for smoother slack distribution across instructions (circuit level), multi-cycle execution of slow instructions (architecture level) and code replacement (compiler level). Our experimental results show that this approach achieves significant energy (34%) and performance (19%) improvement while providing orders of magnitude reduction of timing failure rate.

The proposed functional unit partitioning approach improves the energy efficiency and reliability of functional units by exploiting fine-grained power-gating. For this purpose, a large functional unit like an ALU is partitioned into several smaller (and faster) units based on the instruction usage pattern of the running applications and inherent similarity of the instructions. As a result, the smaller functional units can be power-gated whenever they are not used for a long time. Our simulation results show that the energy efficiency of an ALU can be improved by up to 43.4% in the NTV region. Additionally, the performance can be improved by at least 7.0%, or the reliability can be improved by 11.5 times in the NTV region.

Therefore, by revisiting the design of functional units as important components of data paths and utilizing cross-layer approaches, it is possible to optimize the energy-efficiency, performance, and reliability of NTC designs.

Author Contributions: Conceptualization and methodology, M.S.G. and M.T. All authors have read and agreed to the published version of the manuscript.

Funding: This research received no external funding.

Conflicts of Interest: The authors declare no conflict of interest.

References

1. Moore, G. Cramming more components onto integrated circuits. *Electronics* **1965**, *38*, 114–117. [CrossRef]
2. Moore, G.E. Progress in digital integrated electronics. In Proceedings of the International Electron Devices Meeting (IEDM), Washington, DC, USA, 1–3 December 1975; Volume 21, pp. 11–13.
3. Brock, D.C.; Moore, G.E. *Understanding Moore's Law: Four Decades of Innovation*; Chemical Heritage Foundation: Philadelphia, PA, USA, 2006.
4. Borkar, S. Design challenges of technology scaling. *IEEE Micro* **1999**, *19*, 23–29. [CrossRef]
5. Bohr, M. The new era of scaling in an SoC world. In Proceedings of the International Solid-State Circuits Conference (ISSCC), San Francisco, CA, USA, 8–12 February 2009; pp. 23–28.
6. International Technology Roadmap for Semiconductors (ITRS). Available online: http://www.itrs2.net (accessed on 17 August 2020).
7. International Roadmap for Devices and Systems (IRDS). Available online: https://irds.ieee.org (accessed on 17 August 2020).
8. Dennard, R.H.; Gaensslen, F.H.; Rideout, V.L.; Bassous, E.; LeBlanc, A.R. Design of ion-implanted MOSFET's with very small physical dimensions. *IEEE J. Solid State Circuits* **1974**, *9*, 256–268. [CrossRef]
9. Bohr, M. A 30 year retrospective on Dennard's MOSFET scaling paper. *IEEE Solid State Circuits Soc. Newsl.* **2007**, *12*, 11–13. [CrossRef]
10. Borkar, S.; Chien, A.A. The Future of Microprocessors. *Commun. ACM* **2011**, *54*, 67–77. [CrossRef]
11. Esmaeilzadeh, H.; Blem, E.; Amant, R.S.; Sankaralingam, K.; Burger, D. Dark silicon and the end of multicore scaling. In Proceedings of the International Symposium on Computer Architecture (ISCA), San Jose, CA, USA, 4–8 June 2011; pp. 365–376.
12. Chang, L.; Frank, D.J.; Montoye, R.K.; Koester, S.J.; Ji, B.L.; Coteus, P.W.; Dennard, R.H.; Haensch, W. Practical Strategies for Power-Efficient Computing Technologies. *Proc. IEEE* **2010**, *98*, 215–236. [CrossRef]
13. Esmaeilzadeh, H.; Blem, E.; St. Amant, R.; Sankaralingam, K.; Burger, D. Power Limitations and Dark Silicon Challenge the Future of Multicore. *ACM Trans. Comput. Syst. (TOCS)* **2012**, *30*, 1–27. [CrossRef]
14. Evans, D. The Internet of Things: How the Next Evolution of the Internet Is Changing Everything. In *CISCO White Paper*; Cisco IBSG: San Jose, CA, USA, 2011; pp. 1–11.

15. Ericsson, A. Enabling the internet of things. In *Ericsson Mobility Report: On the Pulse of the Networked Society*; Ericsson AB: Stockholm, Sweden, 2015; p. 10.

16. Middleton, P. Forecast Analysis: Internet of Things—Endpoints, Worldwide, 2016 Update. 2017. Available online: https://www.gartner.com/en/documents/3841268/forecast-analysis-internet-of-things-endpoints-worldwide (accessed on 2 November 2020).

17. Avgerinou, M.; Bertoldi, P.; Castellazzi, L. Trends in data centre energy consumption under the European code of conduct for data centre energy efficiency. *Energies* **2017**, *10*, 1470. [CrossRef]

18. Pedram, M. Power minimization in IC design: Principles and applications. *ACM TRansactions Des. Autom. Electron. Syst. (TODAES)* **1996**, *1*, 3–56. [CrossRef]

19. Pedram, M. Low power design methodologies and techniques: An overview. *Microprocess. Rep.* **1999**, *486*, 66.

20. Tiwari, V.; Singh, D.; Rajgopal, S.; Mehta, G.; Patel, R.; Baez, F. Reducing Power in High-performance Microprocessors. In Proceedings of the Design Automation Conference (DAC), San Francisco, CA, USA, 15–19 June 1998; pp. 732–737.

21. Semeraro, G.; Magklis, G.; Balasubramonian, R.; Albonesi, D.H.; Dwarkadas, S.; Scott, M.L. Energy-efficient processor design using multiple clock domains with dynamic voltage and frequency scaling. In Proceedings of the International Symposium on High-Performance Computer Architecture (HPCA), Cambridge, MA, USA, 2–6 February 2002; pp. 29–40.

22. Burd, T.D.; Pering, T.A.; Stratakos, A.J.; Brodersen, R.W. A dynamic voltage scaled microprocessor system. *IEEE J. Solid State Circuits* **2000**, *35*, 1571–1580. [CrossRef]

23. Chandrakasan, A.P.; Sheng, S.; Brodersen, R.W. Low-power CMOS digital design. *IEEE J. Solid State Circuits* **1992**, *27*, 473–484. [CrossRef]

24. Soeleman, H.; Roy, K.; Paul, B.C. Robust subthreshold logic for ultra-low power operation. *IEEE Trans. Very Large Scale Integr. (VLSI) Syst.* **2001**, *9*, 90–99. [CrossRef]

25. Gonzalez, R.; Gordon, B.M.; Horowitz, M.A. Supply and threshold voltage scaling for low power CMOS. *IEEE J. Solid State Circuits* **1997**, *32*, 1210–1216. [CrossRef]

26. Jain, S.; Khare, S.; Yada, S.; Ambili, V.; Salihundam, P.; Ramani, S.; Muthukumar, S.; Srinivasan, M.; Kumar, A.; Gb, S.K.; et al. A 280 mV-to-1.2 V wide-operating-range IA-32 processor in 32 nm CMOS. In Proceedings of the International Solid-State Circuits Conference (ISSCC), San Francisco, CA, USA, 19–23 February 2012; pp. 66–68.

27. Dreslinski, R.G.; Wieckowski, M.; Blaauw, D.; Sylvester, D.; Mudge, T. Near-threshold computing: Reclaiming moore's law through energy efficient integrated circuits. *Proc. IEEE* **2010**, *98*, 253–266. [CrossRef]

28. Kaul, H.; Anders, M.A.; Mathew, S.K.; Hsu, S.K.; Agarwal, A.; Krishnamurthy, R.K.; Borkar, S. A 320 mv 56 μw 411 gops/watt ultra-low voltage motion estimation accelerator in 65 nm cmos. *IEEE J. Solid State Circuits* **2009**, *44*, 107–114. [CrossRef]

29. Pahlevan, A.; Picorel, J.; Zarandi, A.P.; Rossi, D.; Zapater, M.; Bartolini, A.; Del Valle, P.G.; Atienza, D.; Benini, L.; Falsafi, B. Towards near-threshold server processors. In Proceedings of the Design, Automation & Test in Europe Conference (DATE), Dresden, Germany, 14–18 March 2016; pp. 7–12.

30. Dreslinski, R., Jr. Near Threshold Computing: From Single Core to Many-Core Energy Efficient Architectures. Ph.D. Thesis, University of Michigan, Ann Arbor, MI, USA, 2011.

31. De, V. Fine-grain power management in manycore processor and System-on-Chip (SoC) designs. In Proceedings of the International Conference on Computer-Aided Design (ICCAD), Austin, TX, USA, 2–6 November 2015; pp. 159–164.

32. Zhai, B.; Dreslinski, R.G.; Blaauw, D.; Mudge, T.; Sylvester, D. Energy Efficient Near-threshold Chip Multi-processing. In Proceedings of the International Symposium on Low Power Electronics and Design (ISLPED), Portland, OR, USA, 27–29 August 2007; pp. 32–37.

33. Kaul, H.; Anders, M.; Hsu, S.; Agarwal, A.; Krishnamurthy, R.; Borkar, S. Near-threshold voltage (NTV) design: Opportunities and challenges. In Proceedings of the Design Automation Conference (DAC), San Francisco, CA, USA, 3–7 June 2012; pp. 1153–1158.

34. Alidina, M.; Monteiro, J.; Devadas, S.; Ghosh, A.; Papaefthymiou, M. Precomputation-based sequential logic optimization for low power. *IEEE Trans. Very Large Scale Integr. (VLSI) Syst.* **1994**, *2*, 426–436. [CrossRef]

35. Benini, L.; De Micheli, G.; Macii, E. Designing low-power circuits: Practical recipes. *IEEE Circuits Syst. Mag.* **2001**, *1*, 6–25. [CrossRef]
36. Wei, L.; Chen, Z.; Johnson, M.; Roy, K.; De, V. Design and optimization of low voltage high performance dual threshold CMOS circuits. In Proceedings of the Design Automation Conference (DAC), San Francisco, CA, USA, 15–19 June 1998; pp. 489–494.
37. Fujiwara, F.B. A Neutral Netlist of 10 Combinational Benchmark Circuits. In Proceedings of the International Symposium on Circuits and Systems (ISCAS), Kyoto, Japan, 5–7 June 1985; pp. 695–698.
38. Brglez, F.; Pownall, P.; Hum, R. Accelerated ATPG and fault grading via testability analysis. In Proceedings of the International Symposium on Circuits and Systems (ISCAS), Kyoto, Japan, 5–7 June 1985; pp. 695–698.
39. Sakurai, T.; Newton, A.R. Alpha-power law MOSFET model and its applications to CMOS inverter delay and other formulas. *IEEE J. Solid State Circuits* **1990**, *25*, 584–594. [CrossRef]
40. Markovic, D.; Wang, C.C.; Alarcon, L.P.; Liu, T.; Rabaey, J.M. Ultralow-Power Design in Near-Threshold Region. *Proc. IEEE* **2010**, *98*, 237–252. [CrossRef]
41. Enz, C.C.; Krummenacher, F.; Vittoz, E.A. An analytical MOS transistor model valid in all regions of operation and dedicated to low-voltage and low-current applications. *Analog. Integr. Circuits Signal Process.* **1995**, *8*, 83–114. [CrossRef]
42. Bucher, M.; Bazigos, A.; Krummenacher, F.; Sallese, J.M.; Enz, C. EKV3. 0: An advanced charge based MOS transistor model. A design-oriented MOS transistor compact model. In *Transistor Level Modeling for Analog/RF IC Design*; Springer: Berlin/Heidelberg, Germany, 2006; pp. 67–95.
43. Keller, S.; Harris, D.M.; Martin, A.J. A compact transregional model for digital CMOS circuits operating near threshold. *IEEE Trans. Very Large Scale Integr. (VLSI) Syst.* **2014**, *22*, 2041–2053. [CrossRef]
44. Drego, N.; Chandrakasan, A.; Boning, D. Lack of spatial correlation in MOSFET threshold voltage variation and implications for voltage scaling. *IEEE Trans. Semicond. Manuf.* **2009**, *22*, 245–255. [CrossRef]
45. Hanson, S.; Zhai, B.; Bernstein, K.; Blaauw, D.; Bryant, A.; Chang, L.; Das, K.K.; Haensch, W.; Nowak, E.J.; Sylvester, D.M. Ultralow-voltage, minimum-energy CMOS. *IBM J. Res. Dev.* **2006**, *50*, 469–490. [CrossRef]
46. Kiamehr, S.; Ebrahimi, M.; Golanbari, M.S.; Tahoori, M.B. Temperature-aware dynamic voltage scaling to improve energy efficiency of near-threshold computing. *IEEE Trans. Very Large Scale Integr. (VLSI) Syst.* **2017**, *25*, 2017–2026. [CrossRef]
47. Krishnamurthy, R.K.; Himanshu, K. Ultra-low voltage technologies for energy-efficient special-purpose hardware accelerators. *Intel Technol. J.* **2009**, *13*, 102–117.
48. Bowman, K.; Tschanz, J.; Wilkerson, C.; Lu, S.L.; Karnik, T.; De, V.; Borkar, S. Circuit techniques for dynamic variation tolerance. In Proceedings of the Design Automation Conference (DAC), San Francisco, CA, USA, 26–31 July 2009; pp. 4–7.
49. Ernst, D.; Das, S.; Lee, S.; Blaauw, D.; Austin, T.; Mudge, T.; Kim, N.S.; Flautner, K. Razor: Circuit-level correction of timing errors for low-power operation. *IEEE Micro* **2004**, *24*, 10–20. [CrossRef]
50. Greskamp, B.; Torrellas, J. Paceline: Improving single-thread performance in nanoscale CMPs through core overclocking. In Proceedings of the International Conference on Parallel Architecture and Compilation Techniques (IEEE Computer Society), Brasov, Romania, 15–19 September 2007; pp. 213–224.
51. Ramasubramanian, S.G.; Venkataramani, S.; Parandhaman, A.; Raghunathan, A. Relax-and-retime: A methodology for energy-efficient recovery based design. In Proceedings of the Design Automation Conference (DAC), Austin, TX, USA, 2–6 June 2013; p. 111.
52. Liu, Y.; Ye, R.; Yuan, F.; Kumar, R.; Xu, Q. On logic synthesis for timing speculation. In Proceedings of the International Conference on Computer-Aided Design (ICCAD), San Jose, CA, USA, 5–8 November 2012; pp. 591–596.
53. Golanbari, M.S.; Kiamehr, S.; Tahoori, M.B. Hold-time Violation Analysis and Fixing in Near-Threshold Region. In Proceedings of the International Workshop on Power and Timing Modeling, Optimization and Simulation (PATMOS), Bremen, Germany, 21–23 September 2016.
54. Zhai, B.; Hanson, S.; Blaauw, D.; Sylvester, D. A variation-tolerant sub-200 mV 6-T subthreshold SRAM. *IEEE J. Solid State Circuits* **2008**, *43*, 2338. [CrossRef]

55. Chang, L.; Montoye, R.K.; Nakamura, Y.; Batson, K.A.; Eickemeyer, R.J.; Dennard, R.H.; Haensch, W.; Jamsek, D. An 8T-SRAM for variability tolerance and low-voltage operation in high-performance caches. *IEEE J. Solid State Circuits* **2008**, *43*, 956–963. [CrossRef]

56. Chang, L.; Fried, D.M.; Hergenrother, J.; Sleight, J.W.; Dennard, R.H.; Montoye, R.K.; Sekaric, L.; McNab, S.J.; Topol, A.W.; Adams, C.D.; et al. Stable SRAM cell design for the 32 nm node and beyond. In *Symposium on VLSI Technology*; IEEE: Piscataway, NJ, USA, 2005; pp. 128–129.

57. Kulkarni, J.P.; Kim, K.; Roy, K. A 160 mV, fully differential, robust schmitt trigger based sub-threshold SRAM. In Proceedings of the International Symposium on Low Power Electronics and Design (ISLPED), Portland, OR, USA, 27–29 August 2007; pp. 171–176.

58. Calhoun, B.H.; Chandrakasan, A. A 256kb sub-threshold SRAM in 65nm CMOS. In Proceedings of the International Solid-State Circuits Conference (ISSCC), San Francisco, CA, USA, 6–9 February 2006; pp. 2592–2601.

59. Hu, J.; Yu, X. Low voltage and low power pulse flip-flops in nanometer CMOS processes. *Curr. Nanosci.* **2012**, *8*, 102–107. [CrossRef]

60. Pinckney, N.; Blaauw, D.; Sylvester, D. Low-power near-threshold design: Techniques to improve energy efficiency. *IEEE Solid State Circuits Mag.* **2015**, *7*, 49–57. [CrossRef]

61. Fuketa, H.; Hirairi, K.; Yasufuku, T.; Takamiya, M.; Nomura, M.; Shinohara, H.; Sakurai, T. 12.7-times energy efficiency increase of 16-bit integer unit by power supply voltage (V DD) scaling from 1.2 V to 310 mV enabled by contention-less flip-flops (CLFF) and separated V DD between flip-flops and combinational logics. In Proceedings of the International Symposium on Low Power Electronics and Design (ISLPED), Fukuoka, Japan, 1–3 August 2011; pp. 163–168.

62. Chandra, V.; Aitken, R. Impact of technology and voltage scaling on the soft error susceptibility in nanoscale CMOS. In Proceedings of the IEEE International Symposium on Defect and Fault Tolerance of VLSI Systems, Boston, MA, USA, 1–3 October 2008; pp. 114–122.

63. Cannon, E.H.; Reinhardt, D.D.; Gordon, M.S.; Makowenskyj, P.S. SRAM SER in 90, 130 and 180 nm bulk and SOI technologies. In Proceedings of the IEEE International Reliability Physics Symposium Proceedings (IRPS), Phoenix, AZ, USA, 25–29 April 2004; pp. 300–304.

64. Hazucha, P.; Karnik, T.; Walstra, S.; Bloechel, B.A.; Tschanz, J.W.; Maiz, J.; Soumyanath, K.; Dermer, G.E.; Narendra, S.; De, V.; et al. Measurements and analysis of SER-tolerant latch in a 90-nm dual-V/sub T/CMOS process. *IEEE J. Solid State Circuits* **2004**, *39*, 1536–1543. [CrossRef]

65. Naseer, R.; Draper, J. DF-DICE: A scalable solution for soft error tolerant circuit design. In Proceedings of the International Symposium on Circuits and System (ISCAS), Island of Kos, Greece, 21–24 May 2006; pp. 3890–3893.

66. Zhang, M.; Mitra, S.; Mak, T.; Seifert, N.; Wang, N.J.; Shi, Q.; Kim, K.S.; Shanbhag, N.R.; Patel, S.J. Sequential element design with built-in soft error resilience. *IEEE Trans. Very Large Scale Integr. (VLSI) Syst.* **2006**, *14*, 1368–1378. [CrossRef]

67. Zhou, Q.; Mohanram, K. Transistor sizing for radiation hardening. In Proceedings of the International Reliability Physics Symposium (IRPS), Phoenix, AZ, USA, 25–29 April 2004; pp. 310–315.

68. Ebrahimi, M.; Rao, P.M.B.; Seyyedi, R.; Tahoori, M.B. Low-cost multiple bit upset correction in SRAM-based FPGA configuration frames. *IEEE Trans. Very Large Scale Integr. (VLSI) Syst.* **2016**, *24*, 932–943. [CrossRef]

69. Chang, I.J.; Kim, J.J.; Park, S.P.; Roy, K. A 32 kb 10T sub-threshold SRAM array with bit-interleaving and differential read scheme in 90 nm CMOS. *IEEE J. Solid State Circuits* **2009**, *44*, 650–658. [CrossRef]

70. Fu, X.; Li, T.; Fortes, J.A.B. Soft error vulnerability aware process variation mitigation. In Proceedings of the International Symposium on High Performance Computer Architecture (HPCA), Raleigh, NC, USA, 14–18 February 2009; pp. 93–104.

71. Golanbari, M.S. Design and Optimization for Resilient Energy Efficient Computing. Ph.D. Thesis, Karlsruher Institut für Technologie (KIT), Karlsruhe, Germany, 2019. [CrossRef]

72. Markov, S.; Zain, A.S.M.; Cheng, B.; Asenov, A. Statistical variability in scaled generations of n-channel UTB-FD-SOI MOSFETs under the influence of RDF, LER, OTF and MGG. In Proceedings of the IEEE International SOI Conference (SOI), Napa, CA, USA, 1–4 October 2012; pp. 1–2.

73. Seok, M.; Chen, G.; Hanson, S.; Wieckowski, M.; Blaauw, D.; Sylvester, D. CAS-FEST 2010: Mitigating variability in near-threshold computing. *IEEE J. Emerg. Sel. Top. Circuits Syst.* **2011**, *1*, 42–49. [CrossRef]
74. Hanson, S.; Zhai, B.; Seok, M.; Cline, B.; Zhou, K.; Singhal, M.; Minuth, M.; Olson, J.; Nazhandali, L.; Austin, T.; et al. Performance and variability optimization strategies in a sub-200 mV, 3.5 pJ/inst, 11 nW subthreshold processor. In *IEEE Symposium on VLSI Circuits*; IEEE: Piscataway, NJ, USA, 2007; pp. 152–153.
75. Kakoee, M.R.; Sathanur, A.; Pullini, A.; Huisken, J.; Benini, L. Automatic synthesis of near-threshold circuits with fine-grained performance tunability. In Proceedings of the International Symposium on Low Power Electronics and Design (ISLPED), Austin, TX, USA, 18–20 August 2010; pp. 401–406.
76. Myers, J.; Savanth, A.; Gaddh, R.; Howard, D.; Prabhat, P.; Flynn, D. A subthreshold ARM cortex-M0+ subsystem in 65 nm CMOS for WSN applications with 14 power domains, 10T SRAM, and integrated voltage regulator. *IEEE J. Solid State Circuits* **2016**, *51*, 31–44.
77. Karpuzcu, U.R.; Kim, N.S.; Torrellas, J. Coping with parametric variation at near-threshold voltages. *IEEE Micro* **2013**, *33*, 6–14. [CrossRef]
78. Kim, S.; Kwon, I.; Fick, D.; Kim, M.; Chen, Y.P.; Sylvester, D. Razor-lite: A side-channel error-detection register for timing-margin recovery in 45nm SOI CMOS. In Proceedings of the International Solid-State Circuits Conference (ISSCC), San Francisco, CA, USA, 17–21 February 2013; pp. 264–265.
79. Austin, T.; Bertacco, V.; Blaauw, D.; Mudge, T. Opportunities and challenges for better than worst-case design. In Proceedings of the Asia and South Pacific Design Automation Conference (ASP-DAC), Shanghai, China, 18–21 January 2005; pp. 2–7.
80. Gebregiorgis, A.; Bishnoi, R.; Tahoori, M.B. A Comprehensive Reliability Analysis Framework for NTC Caches: A System to Device Approach. *IEEE Trans. Comput. Aided Des. Integr. Circuits Syst.* **2018**, *38*, 439–452. [CrossRef]
81. Zhai, B.; Pant, S.; Nazhandali, L.; Hanson, S.; Olson, J.; Reeves, A.; Minuth, M.; Helfand, R.; Austin, T.; Sylvester, D.; et al. Energy-efficient subthreshold processor design. *IEEE Trans. Very Large Scale Integr. (VLSI) Syst.* **2009**, *17*, 1127–1137. [CrossRef]
82. Gebregiorgis, A.; Golanbari, M.S.; Kiamehr, S.; Oboril, F.; Tahoori, M.B. Maximizing Energy Efficiency in NTC by Variation-Aware Microprocessor Pipeline Optimization. In Proceedings of the International Symposium on Low Power Electronics and Design (ISLPED), San Francisco, CA, USA, 8–10 August 2016; pp. 272–277.
83. Zhai, B.; Hanson, S.; Blaauw, D.; Sylvester, D. Analysis and mitigation of variability in subthreshold design. In Proceedings of the International Symposium on Low Power Electronics and Design (ISLPED), San Diego, CA, USA, 8–10 August 2005; pp. 20–25.
84. Gebregiorgis, A.; Tahoori, M.B. Fine-Grained Energy-Constrained Microprocessor Pipeline Design. *IEEE Trans. Very Large Scale Integr. (VLSI) Syst.* **2018**, *26*, 457–469. [CrossRef]
85. Nazhandali, L.; Zhai, B.; Olson, J.; Reeves, A.; Minuth, M.; Helfand, R.; Pant, S.; Austin, T.; Blaauw, D. Energy optimization of subthreshold-voltage sensor network processors. *ACM Sigarch Comput. Archit. News* **2005**, *33*, 197–207. [CrossRef]
86. Ramadass, Y.K.; Chandrakasan, A.P. Minimum energy tracking loop with embedded DC–DC converter enabling ultra-low-voltage operation down to 250 mV in 65 nm CMOS. *IEEE J. Solid State Circuits* **2008**, *43*, 256–265. [CrossRef]
87. Ikenaga, Y.; Nomura, M.; Nakazawa, Y.; Hagihara, Y. A circuit for determining the optimal supply voltage to minimize energy consumption in LSI circuit operations. *IEEE J. Solid State Circuits* **2008**, *43*, 911–918. [CrossRef]
88. Koskinen, L.; Hiienkari, M.; Mäkipää, J.; Turnquist, M.J. Implementing minimum-energy-point systems with adaptive logic. *IEEE Trans. Very Large Scale Integr. (VLSI) Syst.* **2016**, *24*, 1247–1256. [CrossRef]
89. Mehta, N.; Makinwa, K.A. Minimum energy point tracking for sub-threshold digital CMOS circuits using an in-situ energy sensor. In Proceedings of the International Symposium on Circuits and Systems (ISCAS), Beijing, China, 19–23 May 2013; pp. 570–573.
90. Mehta, N.; Amrutur, B. Dynamic supply and threshold voltage scaling for CMOS digital circuits using in-situ power monitor. *IEEE Trans. Very Large Scale Integr. (VLSI) Syst.* **2012**, *20*, 892–901. [CrossRef]

91. Wang, A.; Chandrakasan, A.P.; Kosonocky, S.V. Optimal supply and threshold scaling for subthreshold CMOS circuits. In Proceedings of the IEEE Computer Society Annual Symposium on VLSI, Pittsburgh, PA, USA, 25–26 April 2002; pp. 7–11.

92. Calhoun, B.H.; Wang, A.; Chandrakasan, A. Modeling and sizing for minimum energy operation in subthreshold circuits. *IEEE J. Solid State Circuits* **2005**, *40*, 1778–1786. [CrossRef]

93. Karpuzcu, U.R.; Sinkar, A.; Kim, N.S.; Torrellas, J. EnergySmart: Toward energy-efficient manycores for near-threshold computing. In Proceedings of the International Symposium on High-Performance Computer Architecture (HPCA), Shenzhen, China, 23–27 February 2013; pp. 542–553.

94. Seo, S.; Dreslinski, R.G.; Woh, M.; Park, Y.; Charkrabari, C.; Mahlke, S.; Blaauw, D.; Mudge, T. Process variation in near-threshold wide SIMD architectures. In Proceedings of the Design Automation Conference (DAC), San Francisco, CA, USA 3–7 June 2012; pp. 980–987.

95. Wieckowski, M.; Park, Y.M.; Tokunaga, C.; Kim, D.W.; Foo, Z.; Sylvester, D.; Blaauw, D. Timing yield enhancement through soft edge flip-flop based design. In Proceedings of the Custom Integrated Circuits Conference (CICC), San Jose, CA, USA, 21–24 September 2008; pp. 543–546.

96. Joshi, V.; Blaauw, D.; Sylvester, D. Soft-edge flip-flops for improved timing yield: Design and optimization. In Proceedings of the International Conference on Computer-Aided Design (ICCAD), San Jose, CA, USA, 5–8 November 2007; pp. 667–673.

97. Gammie, G.; Wang, A.; Chau, M.; Gururajarao, S.; Pitts, R.; Jumel, F.; Engel, S.; Royannez, P.; Lagerquist, R.; Mair, H.; et al. A 45 nm 3.5 g baseband-and-multimedia application processor using adaptive body-bias and ultra-low-power techniques. In Proceedings of the International Solid-State Circuits Conference (ISSCC), San Francisco, CA, USA, 3–7 February 2008; pp. 258–611.

98. Golanbari, M.S.; Kiamehr, S.; Ebrahimi, M.; Tahoori, M.B. Variation-aware near threshold circuit synthesis. In Proceedings of the Design, Automation & Test in Europe Conference (DATE), Dresden, Germany, 14–18 March 2016.

99. Krimer, E.; Pawlowski, R.; Erez, M.; Chiang, P. Synctium: A near-threshold stream processor for energy-constrained parallel applications. *IEEE Comput. Archit. Lett.* **2010**, *9*, 21–24. [CrossRef]

100. Oboril, F.; Firouzi, F.; Kiamehr, S.; Tahoori, M.B. Negative bias temperature instability-aware instruction scheduling: A cross-layer approach. *J. Low Power Electron.* **2013**, *9*, 389–402. [CrossRef]

101. Oboril, F.; Firouzi, F.; Kiamehr, S.; Tahoori, M. Reducing NBTI-induced processor wearout by exploiting the timing slack of instructions. In Proceedings of the Eighth IEEE/ACM/IFIP International Conference on Hardware/Software Codesign and System Synthesis, Tampere, Finland, 7–12 October 2012; pp. 443–452.

102. Annavaram, M. A case for guarded power gating for multi-core processors. In Proceedings of the International Symposium on High-Performance Computer Architecture (HPCA), San Antonio, TX, USA, 12–16 February 2011.

103. Leverich, J.; Monchiero, M.; Talwar, V.; Ranganathan, P.; Kozyrakis, C. Power management of datacenter workloads using per-core power gating. *Comput. Archit. Lett.* **2009**, *8*, 48–51. [CrossRef]

104. Bose, P.; Buyuktosunoglu, A.; Darringer, J.A.; Gupta, M.S.; Healy, M.B.; Jacobson, H.; Nair, I.; Rivers, J.A.; Shin, J.; Vega, A.; et al. Power management of multi-core chips: Challenges and pitfalls. In Proceedings of the Design, Automation & Test in Europe Conference (DATE), Dresden, Germany, 12–16 March 2012.

105. Hu, Z.; Buyuktosunoglu, A.; Srinivasan, V.; Zyuban, V.; Jacobson, H.; Bose, P. Microarchitectural techniques for power gating of execution units. In Proceedings of the International Symposium on Low Power Electronics and Design (ISLPED), Newport Beach, CA, USA, 9–11 August 2004; pp. 32–37.

106. Golanbari, M.S.; Gebregiorgis, A.; Oboril, F.; Kiamehr, S.; Tahoori, M.B. A Cross-Layer Approach for Resiliency and Energy Efficiency in Near Threshold Computing. In Proceedings of the International Conference on Computer-Aided Design (ICCAD), Austin, TX, USA, 7–10 November 2016.

107. Golanbari, M.S.; Gebregiorgis, A.; Moradi, E.; Kiamehr, S.; Tahoori, M.B. Balancing resiliency and energy efficiency of functional units in ultra-low power systems. In Proceedings of the Asia and South Pacific Design Automation Conference (ASP-DAC), Jeju Island, Korea, 22–25 January 2018.

108. Kaufman, L.; Rousseeuw, P.J. *Finding Groups in Data: An Introduction to Cluster Analysis*; John Wiley & Sons: Hoboken, NJ, USA, 2009; Volume 344.

109. Wang, N.; Wang, H.; Jin, Y.; Ye, J. Integrating operation scheduling and binding for functional unit power-gating in high-level synthesis. In Proceedings of the 2017 IEEE 12th International Conference on ASIC (ASICON), Guiyang, China, 25–28 October 2017; pp. 1129–1132.

110. Jeon, D.; Seok, M.; Chakrabarti, C.; Blaauw, D.; Sylvester, D. A super-pipelined energy efficient subthreshold 240 MS/s FFT core in 65 nm CMOS. *IEEE J. Solid State Circuits* **2012**, *47*, 23–34. [CrossRef]

111. DeGroot, M.H.; Schervish, M.J. *Probability and Statistics*; Pearson Education: London, UK, 2012.

112. Kuhn, K.J.; Giles, M.D.; Becher, D.; Kolar, P.; Kornfeld, A.; Kotlyar, R.; Ma, S.T.; Maheshwari, A.; Mudanai, S. Process technology variation. *IEEE Trans. Electron Devices* **2011**, *58*, 2197–2208. [CrossRef]

113. Wang, N.J.; Quek, J.; Rafacz, T.M.; Patel, S.J. Characterizing the effects of transient faults on a high-performance processor pipeline. In Proceedings of the International Conference on Dependable Systems and Networks, Florence, Italy, 28 June–1 July 2004; pp. 61–70.

114. Synopsys Design Compiler. Available online: https://www.synopsys.com (accessed on 2 November 2020).

115. Cadence Encounter Timing System. Available online: http://www.cadence.com (accessed on 2 November 2020).

116. ECSM Specification. Available online: https://www.si2.org (accessed on 2 November 2020).

117. Sinha, S.; Yeric, G.; Chandra, V.; Cline, B.; Cao, Y. Exploring sub-20nm FinFET design with predictive technology models. In Proceedings of the Design Automation Conference (DAC), San Francisco, CA, USA, 3–7 June 2012; pp. 283–288.

118. Synopsys 32/28 nm PDK. Available online: https://www.synopsys.com/community/university-program/teaching-resources.html (accessed on 2 November 2020).

119. Cadence Virtuoso Variety Statistical Characterization Solution. Available online: http://www.cadence.com (accessed on 2 November 2020).

120. Binkert, N.; Beckmann, B.; Black, G.; Reinhardt, S.K.; Saidi, A.; Basu, A.; Hestness, J.; Hower, D.R.; Krishna, T.; Sardashti, S.; et al. The gem5 simulator. *ACM Sigarch Comput. Archit. News* **2011**, *39*, 1–7. [CrossRef]

121. Hinton, G.; Sager, D.; Upton, M.; Boggs, D.; Group, D.P.; Corp, I. The Microarchitecture of the Pentium 4 Processor. *Intel Technol. J.* **2001**, *1*, 2001.

122. Tan, P.N.; Steinbach, M.; Kumar, V. *Introduction to Data Mining*; Pearson Education: Delhi, India, 2007.

123. Karunaratne, M.; Oomann, B. An optimal memory BISR implementation. *J. Comput.* **2013**, *8*, [CrossRef]

Journal of
*Low Power Electronics
and Applications*

Article

Challenges and Opportunities in Near-Threshold DNN Accelerators around Timing Errors

Pramesh Pandey *,†, **Noel Daniel Gundi** *,†, **Prabal Basu, Tahmoures Shabanian, Mitchell Craig Patrick, Koushik Chakraborty and Sanghamitra Roy**

Bridge Lab, Electrical and Computer Engineering, Utah State University, Logan, UT 84321, USA; prabalb@aggiemail.usu.edu (P.B.); tahmoures@aggiemail.usu.edu (T.S.); mpatrick26@aggiemail.usu.edu (M.C.P.); koushik.chakraborty@usu.edu (K.C.); sanghamitra.roy@usu.edu (S.R.)
* Correspondence: pandey.pramesh1@aggiemail.usu.edu (P.P.); noeldaniel@aggiemail.usu.edu (N.D.G.)
† These authors contributed equally to this work.

Received: 27 August 2020; Accepted: 7 October 2020; Published: 16 October 2020

Abstract: AI evolution is accelerating and Deep Neural Network (DNN) inference accelerators are at the forefront of ad hoc architectures that are evolving to support the immense throughput required for AI computation. However, much more energy efficient design paradigms are inevitable to realize the complete potential of AI evolution and curtail energy consumption. The Near-Threshold Computing (NTC) design paradigm can serve as the best candidate for providing the required energy efficiency. However, NTC operation is plagued with ample performance and reliability concerns arising from the timing errors. In this paper, we dive deep into DNN architecture to uncover some unique challenges and opportunities for operation in the NTC paradigm. By performing rigorous simulations in TPU systolic array, we reveal the severity of timing errors and its impact on inference accuracy at NTC. We analyze various attributes—such as data–delay relationship, delay disparity within arithmetic units, utilization pattern, hardware homogeneity, workload characteristics—and uncover unique localized and global techniques to deal with the timing errors in NTC.

Keywords: near-threshold computing (NTC); deep neural network (DNN); accelerators; timing error; AI; tensor processing unit (TPU); multiply and accumulate (MAC); energy efficiency

1. Introduction

The proliferation of artificial intelligence (AI) is predicted to contribute up to \$15.7 trillion to the global economy by 2030 [1]. To accommodate the AI evolution, the computing industry has already shifted gears to the use of specialized domain-specific AI accelerators along with major improvements in cost, energy and performance. Conventional CPUs and GPUs are no longer able to match up the required throughput and they incur wasteful power consumption through their Von-Neumann bottlenecks [2–6]. However, the upsurge in AI is bound to cater to a huge rise in energy consumption to facilitate power requirements throughout the wide spectrum of AI processing in the cloud data centers to smart edge devices. Andrae et al. have projected that 3–13% of global electricity in 2030 will be used by datacenters [7]. We need to develop energy efficient solutions to drag down this global consumption as low as possible to facilitate the rapid rise in AI compute infrastructure. Moreover, the AI services are propagating deeper into the edge and Internet of Things (IoT) paradigms. Edge AI architectures have several advantages, such as low latency, high privacy, more robustness and efficient use of network bandwidth, which can be useful in diverse applications, such as robotics, mobile devices, smart transportation, healthcare service, wearables, smart speakers, biometric security and so on. The penetration towards the edge will, however, be hit by walls of energy efficiency, as the availability of power from the grid will be replaced by limited power from smaller batteries. In addition, the collective power consumption from the edge and IoT devices will increase drastically. Hence, ultra low power paradigms for AI compute infrastructure, both at server and edge, are inevitable for realizing the complete potential of AI evolution.

Near-threshold computing (NTC) has been a promising roadmap for quadratic energy savings in computing architectures, where the device is operated close to threshold voltage of transistors. Researchers have explored the application of NTC to the conventional CPU/GPU architectures [8–16]. One of the prominent use cases of NTC systems has been to improve the energy efficiency of the computing systems by quadratically reducing the energy consumption per functional unit and increasing the number of functional units (cores). The limit of the general purpose applications to be paralleled [17,18] in this use case has hindered the potential of NTC in CPU/GPU platforms. Deep Neural Network (DNN) accelerators are the representative inference architectures for AI compute infrastructure, which have larger and scalable basic functional units, such as Multiply and Accumulate (MAC) units, which can operate in parallel. Coupled with the highly parallel nature of the DNN workloads to operate on, DNN accelerators serve as excellent candidates for energy efficiency optimization through NTC operation. Substantial energy efficiency can be rendered to inference accelerators in the datacenters. Towards edge, the quadratic reduction in the energy consumption carries the potential to curtail the collective energy demands of the edge devices and makes many compute intensive AI services feasible and practical. Secure and accurate on-device DNN computations can be enabled for AI services, such as face/voice recognition, biometric scanning, object detection, patient monitoring, short term weather prediction and so on at the closest proximity of hardware, physical sensors and body area networks. Entire new classes of edge AI applications delve further into the edge, which are limited by the power consumption from being battery operated and can be enabled by adaptation of NTC.

NTC operation in DNN accelerators is also plagued with almost all the reliability and performance issues as experienced by the conventional architectures. NTC operation is prone to a very high sensitivity to process and environmental variations, resulting in excessive increase in delay and delay variation [14]. This slows down the performance and induces high rate of timing errors in the DNN accelerator. The prevalence, detection and handling of timing error, and its manifestation on the inference accuracy are very challenging for DNN accelerators in comparison to conventional architectures. The unique challenges come from the unique nature of DNN algorithms operating in the complex interleaving among its very large number of dataflow pipelines. Amidst these challenges, the homogeneous repetition of the basic functional units throughout the architecture also bestows unique opportunities to deal with the timing errors. Innovation in the timing properties of a single MAC unit is scalable towards providing a system-wide timing resilience. We observe that very few input combinations to an MAC unit sensitize the delays close to the maximum delay. This provides us with the predictive opportunity where we can predict the bad input combinations and deal with them differently. We also establish the statistical disparity in the timing properties of the multiplier and accumulator inside a MAC unit and uncover unique opportunities to correct a timing error in a timing window derived from the disparity. Through the study of the utilization pattern of MAC units, we present opportunities for fine tuning the timing resilience approaches.

We introduce the basic architecture of DNN in Section 2 with the help of Google's DNN accelerator, Tensor Processing Unit (TPU). We uncover the challenges of NTC DNN accelerators around the performance and timing errors in Section 3. We propose the opportunities in NTC DNN accelerators to deal with the timing errors in Section 4, by delving deep into the architectural attributes, such as utilization pattern, hardware homogeneity, sensitization delay profile and DNN workload characteristics. The quantitative illustrations of the challenges and opportunities are done using the methodology described in Section 5. Section 6 surveys the notable works in the literature close to the domain of NTC DNN accelerators. Finally, Section 7 concludes the paper.

2. Background

DNN inference involves repeated matrix multiplication operation with its input (activation) data and pre-trained weights. Conventional general purpose architectures have a limited number of processing elements (cores) to handle the unconventionally large stream of data. Additionally, the Von-Neumann design paradigm forces to have repeated sluggish and power-hungry memory accesses. DNN accelerators aim for better performance in terms of cost, energy and throughput with respect to conventional computing. DNN workload, although very large, can be highly parallel which opens up possibilities for increasing throughput by feeding data parallel to an array of simple Processing Elements (PE). The memory bottlenecks are relaxed by mixture of different techniques, such as keeping the memory close to the PEs, computing in-memory, architecting dataflow for bulk memory access, using advanced memory, data quantization and so on. DNN accelerators utilize the deterministic algorithmic properties of the DNN workload and embrace maximum reuse of the data through a combination of different dataflow schemes, such as Weight Stationary (WS), Output Stationary (OS), No Local Reuse (NLR), and Row Stationary (RS) [6].

WS dataflow is adapted in most of the DNN accelerators to minimize the memory accesses to the main memory by adding a memory element near/on PEs for storing the weights [19–22]. OS dataflow minimizes the reading and writing of the partial sums to/from main memory, by streaming the activation inputs from neighboring PEs, such as in Shidiannao [22,23]. DianNao [24] follows NLR dataflow, which reads input activations and weights from a global buffer, and processes them through PEs with custom adder trees that can complete the accumulation in a single cycle, and the resulting partial sums or output activations are then put back into the global buffer. Eyeriss [19] embraces an RS dataflow, which maximizes the reuse for all types of data, such as activation, weight and partial sums. Advanced memory wise, DianNao uses advanced eDRAM, which provides higher bandwidth and access speed than DRAMs. TPU [22] uses about 28MiB of on-chip SRAM. There are DNN accelerators which bring computation in the memory. The authors in [25] map the MAC operation to SRAM cells directly. ISAAC [26] PRIME [27], and Pipelayer [28] compute dot product operation using an array of memristors. Several architectures ranging from digital to analog and mixed signal implementations are being commercially deployed in the industry to support the upsurging AI-based economy [22,29–31]. Next, we take an example architecture of the leading edge DNN Accelerator, Tensor Processing Unit (TPU) by Google to better understand the internals of a DNN accelerator compute engine. TPU architecture, already being a successful industrial scale accelerator, has also put its prevalence at the edge through edge TPU [32,33]. We will use this architecture to methodologically explore different challenges and opportunities in an NTC DNN accelerator.

The usage of the systolic array of MAC units, has been recognized as a promising direction to accelerate matrix multiplication operation. TPU employs a 256×256 systolic array of specialized Multiply and Accumulate (MAC) units, as shown in Figure 1 [22]. These MACs, in unison with parallel operation, multiply the 8-bit integer precision weight matrix with the activation (also referred to as input) matrix. Rather than storing to and fetching from the memory for each step of matrix multiplication, the activations stream from a fast SRAM-based activation memory, reaching each column at successive clock cycles. Weights are pre-loaded into the MACs from the weight FIFO ensuring their reuse over a matrix multiplication lifecycle. The partial sums from the rows of MACs move downstream to end up at output accumulator as the result of the matrix multiplication. As a consequence, the Von-Neumann bottlenecks related to memory access are cut down. The ability of the addition of hardware elements to compute the repeating operations not only enables higher data parallelism, but also enables the scaling of the architecture from server to edge applications. This architectural configuration allows TPU to achieve 15–30 X throughput improvement and 30–80 X energy efficiency improvement over the conventional CPU/GPU implementations.

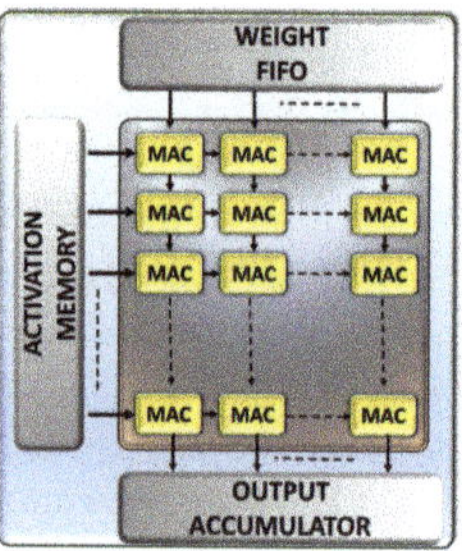

Figure 1. TPU Systolic Array.

3. Challenges for NTC DNN Accelerators

This Section uncovers different challenges in the NTC operation of DNN accelerators. Section 3.1 demonstrates the occurrence of timing errors and their manifestations on the inference accuracy. Section 3.2 presents the challenges in the detection and handling of timing errors in high performance environment essential for a high throughput NTC system.

3.1. Unique Performance Challenge

Even though NTC circuits give a quadratic decrease in energy consumption, they are plagued with severe loss in performance. The loss is general to all the computing architectures as it fundamentally comes from delay experienced by transistors and basic gates. As shown in Figure 2a, basic gates, such as Inverter, Nand and Nor, can experience a delay more than 10×, when operating at near threshold voltages. The gates are simulated in HSPICE for a constant temperature of 25 °C, for more than 10,000 Monte-Carlo iterations. On top of the increase in base delay, the extreme sensitivity of NTC circuits with temperature and circuit noise results in a delay variation of up to 5× [14]. This level of delay increase and delay variability forces the computing architectures to operate in a very relaxed frequency, to ensure the correctness in computation. An attempt to upscale the frequency introduces timing errors. To add to it, the behavior of timing errors is more challenging in DNN accelerators, than conventional CPU/GPU architectures.

In Figure 2b, we plot the rate of timing errors in the inference computations in a TPU of eight DNN datasets using the methodology described in Section 5. As there can be computations happening in all the MAC units in parallel, crossing a delay threshold brings a huge number of timing errors at once. The rate of the timing errors is different for different datasets due to its dependence on the number of actual operations which actually sensitize the delays. As DNN workload consists of several clusters of identical values, (usually zero [2]), DNN workloads tend to decrease the overall sensitization of hardware delays. The curves tend to flatten towards the end as almost all delay sensitizing operations are saturated as timing errors at prior voltages.

Inference Accuracy is the measure of the quality of the DNN applications. In Figure 2c, we show the drop in inference accuracy of MNIST dataset from 98%. We conservatively model the consequence of timing error as the flip of the correct bit with 50% probability in the MAC unit's output. DNN workloads have an inherent algorithmic tolerance to error until a threshold [34,35]. In line with the tolerance, we see that the accuracy variation is maintained under 5% until the rate of timing errors is 0.02%. However, after the error tolerance exceeds this, the accuracy falls very rapidly with a landslide effect. By virtue of the complex interconnected pipelining of the operations, the timing error induced incorrect computations add up rapidly as errant partial sums spread over most parts of the array, towards a bad accuracy. After about 0.045% of timing errors, the accuracy rapidly drops from 84% to a totally unusable inference accuracy of 35% on the timing error difference window of just 0.009% (highlighted in blue color).

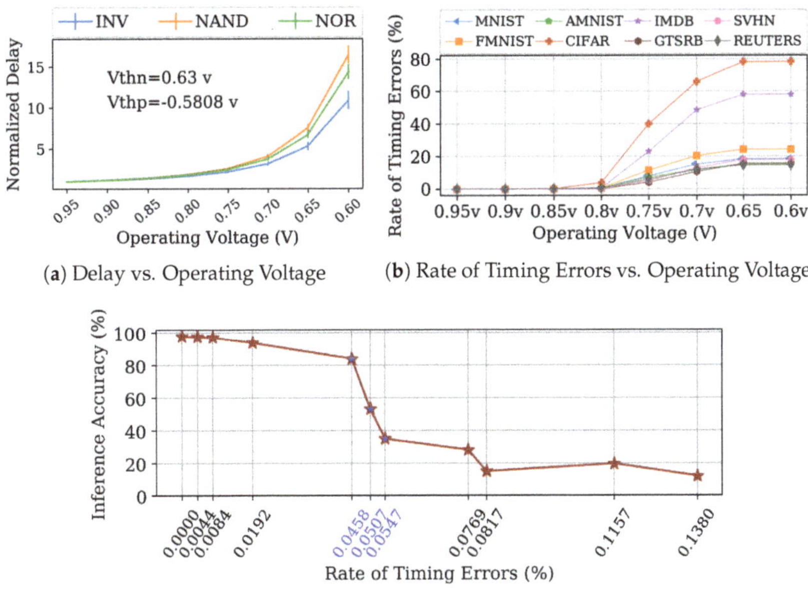

(**a**) Delay vs. Operating Voltage   (**b**) Rate of Timing Errors vs. Operating Voltage

(**c**) Inference Accuracy vs. Rate of timing errors for MNIST dataset

Figure 2. (**a**) shows the increase in base delay with the decrease in operating voltage, (**b**) demonstrates the increase in the rate of timing errors with the decrease in operating voltage and (**c**) shows the effect of timing errors in inference accuracy. (Detailed methodology in Section 5).

This points to a completely impotent DNN accelerator at only a timing error rate of less than 0.06%. This treacherous response of timing errors to inference accuracy in DNN accelerators is magnified at NTC. It further compels the NTC operation to consider all the process, aging and environmental extremes just to prevent a minuscule ($\ll$0.1%), yet catastrophic rate of timing errors, resulting in extremely sluggish accelerators. This creates further distancing in the adaptation of NTC systems into mainstream server/edge applications. Hence, innovative and dynamic techniques to reliably identify and control timing errors are inevitable for NTC DNN accelerators.

3.2. Timing Error Detection and Handling

DNN accelerators have been introduced to offer a throughput, which is difficult to extract from conventional architectures for DNN workload. However, the substantial performance lag at NTC operation hinders the usefulness of NTC DNN accelerators in general. So, in order to embrace NTC design paradigm, the DNN accelerators have to be operated at very tight frequencies, with an expectation and detection of timing errors, followed by their appropriate handling. In this Section, we explore the challenges in the timing error detection and handling for NTC DNN accelerators in these high performance points through the lens of techniques available for conventional architectures.

Razor [36] is one of the most popular timing error detection mechanisms. It detects a timing error by augmenting a shadow flip-flop driven by a delayed clock, to each flip-flop in the design driven by a main clock. Figure 3 shows Razor's mechanism through timing diagrams. A delayed clock can be obtained by a simple inversion of the main clock. Figure 3a depicts the efficient working conditions of a Razor flip-flop. Delayed transitioning of *data2* results in *data1* (erroneous data) being latched onto the output. However, shadow flip-flop detects a detained change in the computational output and alerts the system via an error signal, generated by the comparison of prompt and delayed output. The frequency scaling for very high performances decreases the clock period thereby, diminishing the timing error detection window or speculation window. Shrinking the speculation window, prevents detection of delayed transitions in the computational output and leads to a huge number of timing errors going undiscovered. Figure 3b depicts the late transition in the Razor Input during the second half of the *Cycle 1*. Since the transition occurs after the rising edge of delayed clock, the data manifestation goes undetected by the shadow flip-flop resulting in an undiscoverable timing error. Rapid transition from *data2* to *data3*, and in-time sampling of *data3* at the positive edge of clock during Cycle 2, ushers the complete withdrawal of *data2* during Cycle 1 from the respective MAC computation. Hence, the undiscoverable timing error leads to the usage of *data1* (erroneous data) during Cycle 1, in place of the *data2* (authentic data). Figure 3c demonstrates a delayed transition from *data2* to *data3*, causing *data3* to miss the sampling point (positive edge of the clock in Cycle 2). Shadow flip-flop appropriately procures the delayed data (i.e., *data3*), spawning an architectural replay and delivering *data3* to Razor output during the next operational clock cycle (i.e., Cycle 3). However, authentic data (i.e., *data2*) to be used for MAC computation during Cycle 1, is again ceded from the appropriate MAC's computation. Erroneous values (i.e., *data1*) used during Cycle 1, render to an erroneous input being used in MAC calculations, generating faulty output. Hence, the undiscoverable timing error again leads to an erroneous computation.

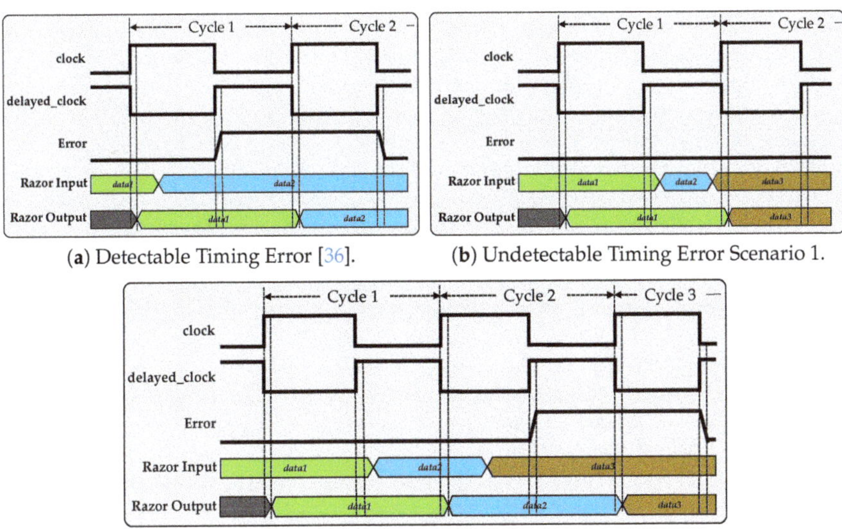

(**a**) Detectable Timing Error [36].

(**b**) Undetectable Timing Error Scenario 1.

(**c**) Undetectable Timing Error Scenario 2.

Figure 3. (**a**) depicts Razor's timing error handling effectiveness in a TPU Systolic Array during a standard detectable timing error occurrence scenario. (**b**,**c**) demonstrate the Razor's limitations in handling undetectable timing errors arising at high performance scaling. In (**c**), even though *data2* is sampled at Cycle 2, architectural replay invoked due to late transition and detection of *data3* ensues the relinquishing of *data2*, entirely.

Figure 4 depicts the undiscoverable timing errors as a percentage composition of the total timing errors at various performance scaling, for different datasets. The composition of undiscoverable timing errors rises linearly until $1.7\times$ the baseline performance. However, with the further increase in performance, the percentage of undiscoverable timing errors grows exponentially, following along with the landslide effect contributed by a large number of parallel operating MACs. This exponential composition of the undetectable timing errors points towards a hard wall of impracticality for razor-based timing error detection approaches.

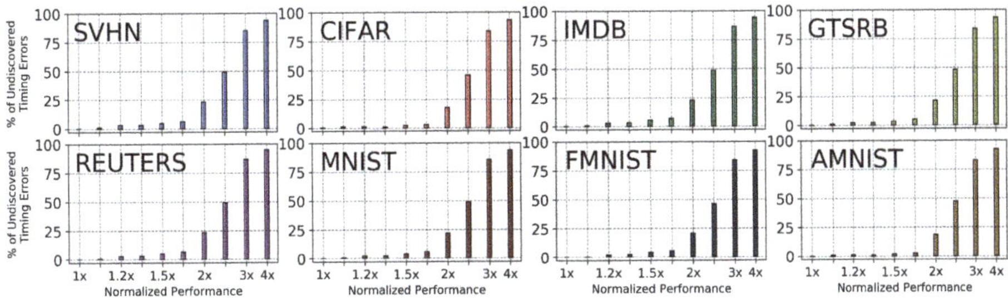

Figure 4. Rate of undiscovered timing errors vs. Performance Scaling.

For handling of timing errors, the architectural replay of the errant computation has been a feasible and preferred way in the conventional general purpose architectures by virtue of their handful of computation pipelines [36,37]. However, the DNN accelerator involves hundreds of parallel computation pipelines and complex and coordinated interleaved dependencies among each other, which in the worst case, forces the sacrifice of computation in all the MACs for correcting just one timing error. For instance, a timing error rate of just 0.02% (starting point of accuracy fall in Figure 2c) for the matrix multiplication of a 256 × 256 activation and weight matrices in a 256 × 256 (N = 256) TPU systolic array, introduce ~3355 timing errors. Distributing the errors to the multiplication life cycle of 766 (3N-2) clock cycles creates approximately four errors per clock cycle. Even with a conservative estimate with a global sacrifice of only one relaxed clock cycle for all the errors per cycle, we get a throughput loss of more than 100%. Scaling to the inflated hardware size at the NTC design paradigm, the throughput coming from a DNN accelerator will be severely undermined by holistic stalling sacrifice of the MAC computations.

Fine tuned distributed techniques with pipelined error detection and handling [37,38] also incur larger overheads than conventional razor-based detecting and handling [35]. A technique designed exclusively for DNN accelerators, TE-Drop [35], skips an errant computation rather than replaying it by exploiting the inherent error resilience of DNN workload. However, the skipping is only triggered by a razor-based detection mechanism and thus can work for razor detectable errors only. With this comprehensive impracticality of the conventional detection and handling of timing errors for NTC performance needs, it calls for further research around the scalable solutions which can provide timing error resilience at vast magnitudes encompassing the entirety of functional units.

4. Opportunities for NTC DNN Accelerators

This Section reveals the unique opportunities for dealing with the timing errors in NTC DNN accelerators. Section 4.1 does an extensive analysis of the delay profile of the accelerators pointing towards predictive opportunities. Section 4.2 presents a novel opportunity of handling a timing error without performance loss. Section 4.3 uncovers the opportunity of an added layer of design intelligence derived from the utilization pattern of MACs in the DNN accelerators.

4.1. Predictive Opportunities

The DNN accelerator's compute unit is comprised of homogeneously repeated simple functional units, such as MAC. In the case of TPU, the multiplier of the MAC unit operates on 8-bit activation and 8-bit weight. Delay is sensitized as a function of the change in these inputs to the multiplier. Weight is stagnant for a computation life cycle and the activations can be changed after each clock cycle. We create an exhaustive set of a 8-bit activation changed to another 8-bit activation for all possible 8-bit weights, leading to 256 × 256 × 256 = 16,777,216 unique combinations. Injecting these entries to our in-house Static Timing Analysis (STA) tool, we plot the exhaustive delay profile of the multiplier in Figure 5 as a histogram. We consider the maximum delay as one clock cycle period, to ensure an error free assessment. The histogram is divided into ten bars, each representing the percentage of the combinations falling into the ranges of clock cycles in the x-axis. We see that more than 95% of the sensitized delays fall in the range of 20–60% of the clock cycle and only about 3% of the exhaustive input sequences incur a delay of more than 60% of the clock period. As these limited sequences are the leading candidates to cause timing error in the NTC DNN accelerator, we see a huge opportunity for providing immense timing error resilience by massively curtailing the potential timing errors through their efficient prediction beforehand.

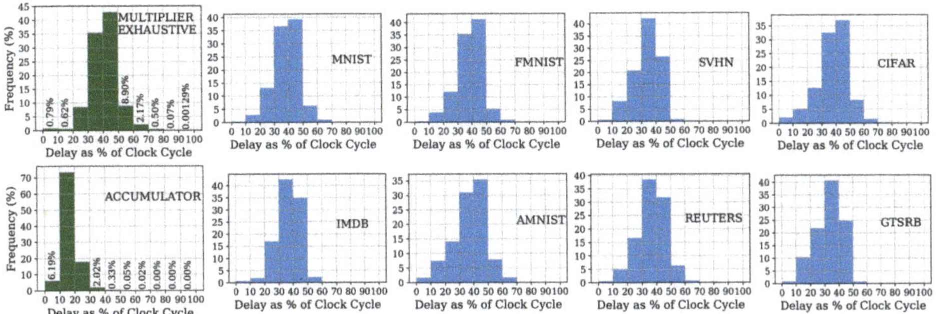

Figure 5. Delay Profiles of multiplier, accumulator and different datasets.

As opposed to exhaustive prediction, the amount of prediction required for achieving timing error resilience is drastically reduced, while operating on real DNN workloads. Real DNN workloads will only have a subset of the exhaustive input combinations. Additionally, real DNN workloads consist of a large share of clustered, non-useful data in the computation, which can be skipped for computation. For instance, test images of MNIST and FMNIST datasets in Figure 6a,b visually show the abundant wasteful black regions in their 28 $\times$ 28 pixels of data, which can be quantized to zero. As the result of multiplication with zero is zero, we can skip the computation altogether by directly passing zero to the output. This technique has been well recognized by researchers to improve the energy efficiency of DNN accelerators, as Zero-Skip [2,19,39]. Figure 6c shows the percentage of the ZS computations in eight DNN datasets that can be transformed to have a close-to-zero sensitization delay. We see that, on average, the DNN datasets can have about ~75% of the ZS computations.

This points to an opportunity of cutting down timing error, causing input combinations that have to be correctly predicted by ~75% in the best case. We plot the delay profile of the datasets for all the input sequences remaining after the massive reduction via ZS in Figure 5. It is evident that there is an ample delay variation across the datasets and the sensitization delays are clustered below 50% of the clock cycle. This further curtails the number of predictions required.

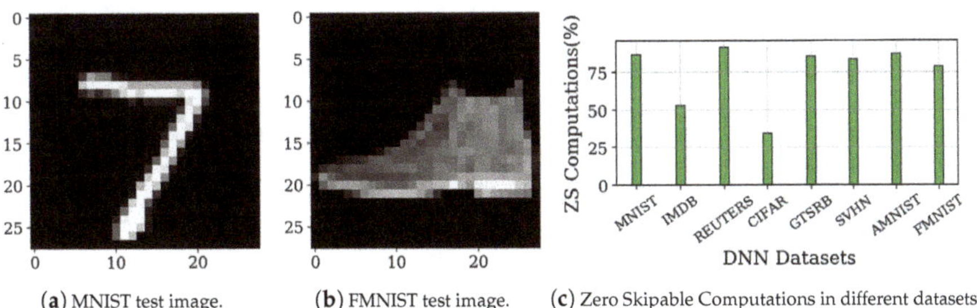

(**a**) MNIST test image.　(**b**) FMNIST test image.　(**c**) Zero Skipable Computations in different datasets.

Figure 6. Demonstration of the share of computations which can be safely skipped. The black region pixels in the MNIST and FMNIST test images can be quantized to zero and all the computations involving those pixels can be "zero skipped".

Hence, given that even a very small rate of timing error ($\ll$0.1%) can cause a devastation in the inference accuracy (Section 2), the absolute feasibility of predictive approach to capture and deal with limited number of timing error causing input sequences in real DNN workloads serves as a boon. Moreover, prediction, along with ZS, provides unique opportunity for performance overscaling of NTC DNN accelerators to match STC performance. Even assuming that all the computations will incur a timing error at such environments, we would only have to predict only ~25% of the computations. We can employ NTC timing error resilience techniques, such as voltage boosting, for only the predicted bad input sequences and then scale it to larger TPU sizes. We have exploited this predictive opportunity and voltage boosting, to enable high performing NTC TPU in [40] with around less than 5% area and power overheads. Prediction can also be used in conjunction or on top of other timing error resilience schemes, such as TE-Drop [35], MATIC [41], FATE [42] and ARES [43], to yield better results. Extended research on the possibility of decreasing the amount of predictions on high performance scaling can be done with inspirations from algorithmic domains, layer wise delay analysis, and so on, to truly boost the adaptation of NTC DNN accelerators.

4.2. Opportunities from Novel Timing Error Handling

In this section, we discuss the opportunities in the handling of timing errors, bestowed by the unique architectural intricacies inside the MAC unit. We start by comparing the delay profiles of the arithmetic units, multiplier and accumulator, inside of the MAC unit. We prepare an exhaustive delay profile for multiplier as described in Section 4.1. As an exhaustive delay profile of accumulator is not feasible with its much larger input bit width (state space), we prepare its delay profile by using the outputs from multiplier fed with exhaustive inputs. From the delay histograms of the multiplier and accumulator in Figure 5, we see that accumulator operation statistically takes much less computation time than the multiplier. Over 97% of the accumulator sensitized delays fall within 30% of the clock cycle. Hence, an MAC operation temporally boils down as an expensive multiplier operation (Section 4.1) followed by a minuscule accumulator operation. This disparate timing characteristic in MAC's arithmetic units, clubbed with the fixed order of operation, opens up a distinct timing error correction window to correct timing violations in a MAC unit, without any performance overhead.

Figure 7 shows a column-wise flow of computation using two MAC units. When MAC 1 encounters a timing error, the accumulator output (*MAC 1 Out*) provided by MAC 1 and the accumulator input (*MAC 2 In*) received by MAC 2 after data synchronization, will be an erroneous value. Although MAC 2 multiplier operation is unaffected by the faulty accumulator input, MAC 2 accumulator output (*MAC 2 Out*), will be corrupted due to the faulty input (*MAC 2 In*). While the MAC 2 accumulator is waiting for the completion of MAC 2 multiplier's operation, the faulty input (*MAC 2 In*) can hypothetically use this extra time window to correct itself. Since the accumulator requires a statistically smaller computation period, the accumulation process with the corrected input can be completed within a given clock period, thereby preserving the throughput of the DNN accelerator. Next, we discuss a simple way of performing this time stealing in hardware through the intelligent remodeling of razor.

Typical use of Razor in a Systolic Array incurs a huge performance overhead due to architectural replay when a timing error is detected (Section 3.2). However, a minimal modification in the Razor design aids in the exploitation of timing error correction window during Systolic Array operation. Figure 8a depicts the change employed in the razor which replays the errant computation [36] to ReModeled razor, which can propagate the authentic value to the downstream MAC without replaying or skipping the computation.

Figure 8b depicts the error mitigation procedure using ReModeled Razor. In ReModeled razor, the timing error manifests as a different (correct) value from a shadow flip-flop, which overrides and propagates over the previous erroneous value passed through the main flip-flop. Since accumulation operation statistically requires much less than 50% of the clock cycle (Figure 5), the accumulator in the downstream MAC can quickly recompute with the correct value. The presence of ZS computations (Figure 6c) further aids the timing error correction window, as the skippable multiplier operations leave the accumulator to sensitize only to the upstream MAC's output. Figure 9 demonstrates the RTL simulations for Razor/Remodeled Razor in the absence and presence of a timing error. Figure 9a depicts the standard waveform for a timing error free operation. Figure 9b shows an occurrence of a timing error due to a minor delay in the Razor input and the stretching of operational cycle to procure the correct value for the re-computation process. Figure 9c elaborates the detection and correction of the timing error, within the same clock cycle. Although, Razor adds buffers on short paths to avoid hold time violations, Zhang et al. [44] have modeled the additional buffers in timing simulations and determined that in a MAC unit only 14 out of 40 flip-flops requires a protection by Razor, which only adds a power penalty of 3.34%. The addition of ReModeled Razor into the Systolic Array adds an area overhead of only 5% into the TPU. In light of these findings, it is evident that we can extract novel opportunities to handle the timing errors at NTC by additional research on fine-tuning the combinational delays among the various arithmetic blocks inside the DNN accelerator. We have exploited this opportunistic timing error mitigation strategy in [45] to provide timing error resilience in a DNN systolic array.

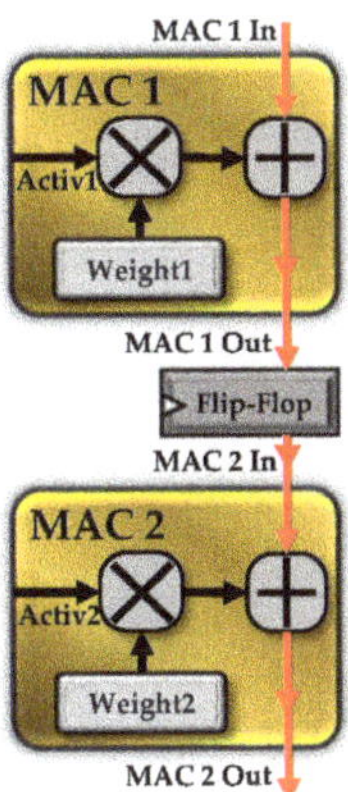

Figure 7. Column-Wise MAC Operation.

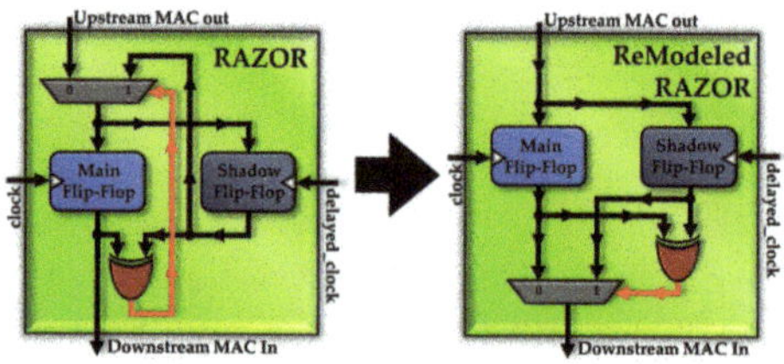

(**a**) Razor [36] to ReModeled Razor.

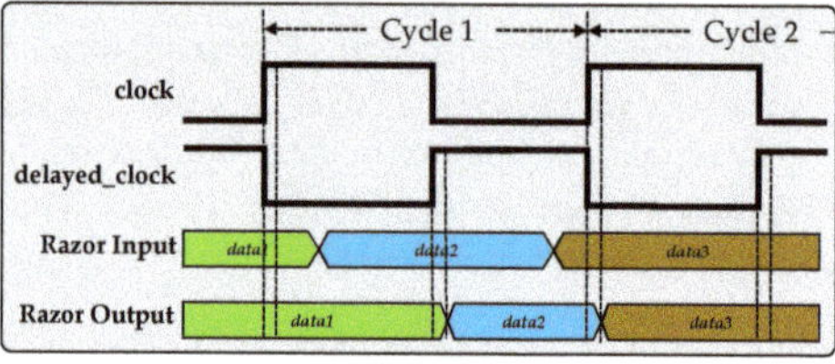

(**b**) Error Correction.

Figure 8. (**a**) depicts the modifications employed in Razor to exploit the timing error correction window. (**b**) demonstrates the error correction process using Remodeled Razor scheme. The error is detected and corrected in the same clock cycle of operation.

(**a**) Error free operation in Razor/ReModeled Razor.

(**b**) Error handled by Razor using instant replay.

(**c**) Error handled by ReModeled Razor without a loss in clock period.

Figure 9. Timing simulations for error-free and error-handling scenarios.

4.3. Opportunities from Hardware Utilization Trend

Although DNN accelerators host very high number of functional units, not all of them are computationally active through out the computation lifecycle. We can leverage the hardware utilization pattern of the DNN accelerators to fine tune our architectural optimization strategies at NTC. In a TPU systolic array, activation is streamed to each row of MACs from left to right with a delay of one clock cycle per row. As the MACs can only operate on the presence of an activation input, a unique utilization pattern of the MAC units is formed. Visual illustration of the computation activity for a scaled down 3×3 systolic array during the multiplication of 3×3 activation and weight matrices is presented in Figure 10. Computationally active and idle MACs are represented by red and green, respectively. Figure 11a plots the computationally active MACs in every clock cycle as a fraction of the total MACs in the systolic array for the actual 256×256 array, for the multiplication of 256×256 weight and activation matrices, corresponding to a batch size of 256. The batch size dictates the amount of data available for continuous streaming, which consequently decides the computation cycles (width of the curve) and maximum usage (maximum height of the curve). Batch sizes are dependent and limited by the hardware resources to hold the activations and AI latency requirements. Regardless of the batch size, the utilization curve follows a static pattern where the activity peaks at the middle of the computation life cycle with a symmetrical rise and fall.

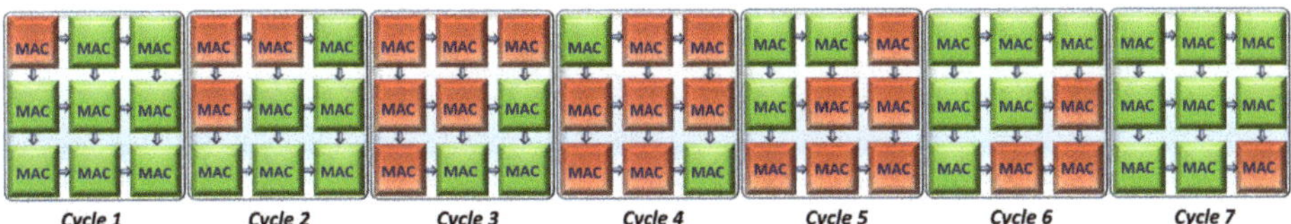

Figure 10. Cycle accurate utilization pattern of the MAC units during the multiplication of 3×3 activation and weight matrices in a 3×3 systolic array.

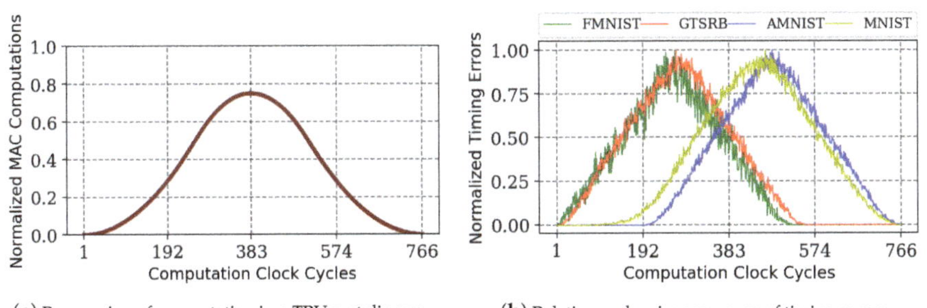

(**a**) Progression of computation in a TPU systolic array.

(**b**) Relative cycle wise occurence of timing errors.

Figure 11. The progression of number of MACs involved in computation guides the trend of timing error over the clock cycles for different datasets.

Timing errors are only encountered when the MACs are actively involved in the computation. This means that the intensity of timing errors in any clock cycle are always capped by the number of computationally active cycles in that cycle and the rate of occurrence of timing errors follow the trend of computation activity. We plot the cycle wise trend of timing errors in Figure 11b for four datasets. It is evident that the progression of timing errors is guided by the trend of compute activity. Any timing error handling schemes used for general purpose NTC paradigm can be further optimized for adaptation to DNN accelerators by leveraging this architectural behavior of timing errors. For instance, aggressive voltage boosting can be implemented at only the windows of maximum timing errors. Additionally, the timing error control schemes can be relaxed temporally at the windows of low probabilities of timing errors. In addition, predictive schemes can be tailored by adjusting the prediction windows according to these static timing error probabilities guided by hardware utilization pattern.

5. Methodology

In this Section, we explain our extensive cross-layer methodology, as depicted in Figure 12, to explore different challenges and opportunities of NTC DNN accelerators.

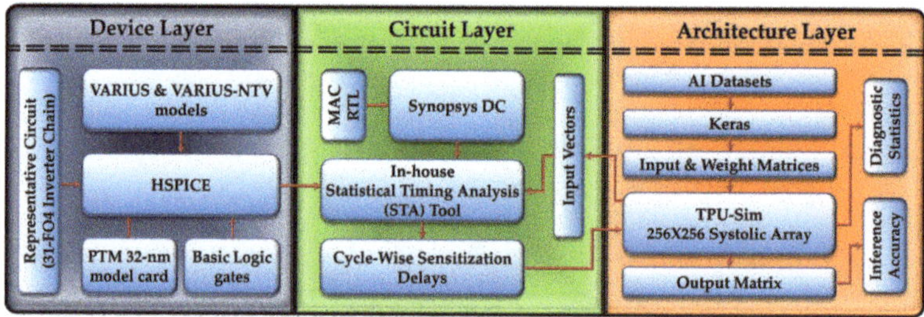

Figure 12. Cross Layer Methodology.

5.1. Device Layer

We simulated basic logic gates (viz., Nand, Nor and Inverter) in HSPICE using basic CMOS 32-nm Predictive Technology Model libraries [46], across the spectrum of supply voltages. We used the 31-stage FO4 inverter chain as a representative of various combinational logics in a TPU for accurate estimation. We incorporated the impact of the PV at NTC using the VARIUS-NTV [47] model. The characteristics of the basic gates were mapped to the circuit layer (Section 5.2), to ascertain the sensitized path delays in a MAC at different voltages.

5.2. Circuit Layer

We developed the Verilog RTL description of MAC unit as the functional element of the systolic array. Our in-house Statistical Timing Analysis (STA) tool takes the synthesized netlist, input vectors for the netlist, and the timing properties of the logic gates. We synthesized the MAC RTL using the Synopsys Design Compiler and Synopsys's generic 32 nm standard cell library, to get the synthesized netlist. The input vectors for the MAC units were the activation, weight and partial sum input, which came from the cycle accurate simulation described in Section 5.3. The changes in timing properties of the logic gates on different operating conditions came from the HSPICE simulation, described in Section 5.1. The STA tool propagated the path in the netlist which was sensitized by the given input vectors and calculated the delay coming from the logic gates in the path by mapping them to the delays from HSPICE. Hence, we got an accurate estimation of delay sensitized by any input change for a range of operating conditions. Our baseline NTC operating voltage and frequency were 0.45 v and 67 MHz.

5.3. Architecture Layer

Based on the architectural description detailed in [22], we developed a cycle-accurate TPU systolic array simulator—TPU-Sim—in C++, to represent a DNN accelerator. We integrated the STA tool

(Section 5.2) with TPU-Sim, to accurately model timing errors in the MACs, based on real data-driven sensitized path delays. We created a real TPU-based inference eco-system by conjoining TPU-Sim with Keras [48] on Tensorflow backend. First, we train several DNN applications (viz., MNIST [49], Reuters [50], CIFAR-10 [51], IMDB [52], SVHN [53], GTSRB [54], FMNIST [55], FSDD [56]). Then, we extracted each layer's activation as input matrices. We extract the trained model weights using Keras built-in functions. As the Keras trained model is at high floating point precision, we processed the inputs and weights into multiple 256×256 8-bit-integer matrices as TPU operates at 8-bit integer precision on a 256×256 systolic array. TPU-Sim is invoked with each pair of these TPU-quantized input and weight matrices. Each MAC operation in the TPU is mapped to a dedicated MAC unit in the TPU-Sim. The delay engine was invoked with each input vector arriving at the MAC to get the assessment of a timing error. The output matrices from the TPU-Sim were combined and compared with the original test output to evaluate the inference accuracy. We paralleled our framework for handling large amounts of test data using Python Multiprocessing.

6. Related Works

Several schemes have been proposed to increase the reliability and efficiency of neural network accelerators. Sections 6.1–6.3 provide brief accounts of enhancement methodologies which are categorized based on the emphasis on the architectural/design components.

6.1. Enhancements around Memory

In this Section, we enlist methodologies which target memory to provide improved performance. Kim et al. [57] demonstrate that a significant accuracy loss is caused by certain bits during faulty DNN operations and using this fault analysis proposed a fault tolerant reliability improvement scheme—DRIS-3—to mitigate the faults during DNN operations. Chandramoorthy et al. [58] present a technique which dynamically boosts the supply voltage of the embedded SRAMs to achieve superior energy savings. Yin et al. [59] evaluate thermal issues in an NN accelerator 3D memory and propose a "3D + 2.5D" integration processor named Parana which integrates 3D memory and the NPU. Parana tackles the thermal problem by lowering the amount of memory access and changing the memory access patterns. Nguyen et al. [60] propose a new memory architecture to adaptively control the DRAM cell refresh rate to store possible errors leading to a reduction in power consumption. Salami et al. [61], based on a thorough analysis of the NN accelerator components devise a strategy to appropriately mask the MSBs, to recover the corrupted bits, thereby enhancing the efficiency by mitigating the faults.

6.2. Enhancements around Architecture

This Section focuses on techniques which have provided enhancements around the architectural flow/components of the DNN Accelerator. Li et al. [62] demonstrate that, by providing appropriate precision and numeric range to values in each layer, reduces the failure rate by $200\times$. In each layer of DNN, this technique uses a "symptom based fault detection" scheme to identify the range of values and adds a 10% guard-band. Libano et al. [63] propose a scheme to design and apply triple modular redundancy selectively to the vulnerable NN layers to effectively mask the faults. Zhang et al. [35] propose a technique, TE-Drop, to tackle timing errors arising due to aggressive voltage scaling. The occurrence of a timing error is detected using Razor flip-flop [36]. The MAC encountering timing error steals a clock cycle from the downstream MAC to recompute the correct output and bypasses the downstream MAC's output with its output. Choi et al. [64] demonstrate a methodology to enhance the resilience of the DNN accelerator based on the sensitivity variations of neurons. The technique detects an error in the multiplier unit by augmenting each MAC unit with a Razor Flip-Flop [36] between multiplier and accumulator unit.

Occurrence of a timing error will be lead to the upstream partial sum to be bypassed on to the downstream MAC unit. Zhang et al. [65] address the reliability concerns due to permanent faults arising from MAC units by mapping and pruning the weights of faulty MAC units.

6.3. Enhancements around Analog/Mixed-Signal Domain

Analog and mixed-signal DNN accelerators are also making a mark in the Neural Network computing realm. Analog and mixed-signal accelerators use enhanced Analog-to-Digital converter (ADC), Digital-to-Analog converter (DAC) for encoding/decoding and Non-Volatile Memories, such as ReRAM's in DNN-based computations. Eshraghian et al. [66] utilize the frequency dependence of v-i place hysteresis to relieve the limitation on the single-bit-per-device and allocating the kernel information to the device conductance and partially to the frequency of the time-varying input. Ghodrati et al. [67] propose a technique BIHIWE, to address the issues in mixed-signal circuitry due to restricted scope of information encoding, noise susceptibility and overheads due to Analog to Digital conversions. BIHIWE, bit-partitions vector dot-product into clusters of low-bitwidth operations executing in parallel and embedding across multiple vector elements. Shafiee et al. [26] demonstrate a scheme ISAAC, by implementing a pipelined architecture with each neural network layer being dedicated specific crossbars and heaping up the data between pipe stages using eDRAM buffers. ISAAC also proposes a novel data encoding technique to reduce the analog-to-digital conversion overheads and performs a design space inspection to obtain a balance between memristor storage/compute, buffers and ADCs on the chip. Mackin et al. [68] propose the usage of crossbar arrays of NVMs to implement MAC operations at the data location and demonstrates simultaneous programming of weights at optimal hardware conditions and exploring its effectiveness under significant NVM variability. These recent developments in Analog and Mixed-signal DNN accelerators envision employing ADC, DAC and ReRAM's in NTC DNN accelerators to yield better energy efficiency. Successive Approximation Register ADCs can be efficiently utilized at NTC owing to their simple architecture and low power usage [69–73]. In addition, the efficacy of ReRAM [74–79] in a low power computing environment provides a promising direction in DNN accelerators venturing into the NTC realm.

7. Conclusions

The NTC design paradigm, being a stronger direction towards energy efficiency, is plagued by timing errors for DNN applications. In this paper, we illustrate the unique challenges coming from the DNN workload and attributes of the occurrence and impact of timing errors. We discover that NTC DNN accelerators are challenged by landslide increases in the rate timing errors and the inherent algorithmic tolerance of DNNs to timing errors is quickly surpassed with a sharp decline in the inference accuracy. We explore the data–delay relationship with the practical DNN datasets to uncover an opportunity of providing bulk timing error resilience through prediction of a small group of high delay input sequences. We explore the timing disparities in the multiplier and accumulator to propose opportunities for providing an elegant timing error correction without performance loss. We correlate the hardware utilization pattern to outshine the broad control strategies for timing error resilience techniques. The application of NTC in the domain of DNN accelerators is a relatively newer domain and this article serves to present concrete possible directions towards timing error resilience, which remains to be a big hurdle for NTC systems. Energy efficient AI evolution is the need of the hour and further research with promising domains, such as NTC is required.

Author Contributions: Conceptualization, P.P. and N.D.G.; methodology, P.P., N.D.G., P.B. and T.S.; software, P.B., P.P., N.D.G. and T.S.; validation, P.P. and N.D.G.; formal analysis, P.P. and N.D.G.; investigation, P.P., N.D.G., P.B., T.S. and M.C.P.; resources, P.P., N.D.G., P.B., T.S., K.C. and S.R.; data curation, P.P.; writing—original draft preparation, P.P. and

N.D.G.; writing—review and editing, P.P., N.D.G., K.C. and S.R.; visualization, P.P. and N.D.G.; supervision, K.C. and S.R.; project administration, P.P.; funding acquisition, K.C. and S.R. All authors have read and agreed to the published version of the manuscript.

Funding: This research was supported in part by National Science Foundation under grant numbers CAREER-1253024 and CNS-1421022.

Conflicts of Interest: The authors declare no conflict of interest.

References

1. AI Will Add 15 Trillion to the World Economy by 2030. Available online: https://www.forbes.com/sites/greatspeculations/2019/02/25/ai-will-add-15-trillion-to-the-world-economy-by-2030/ (accessed on 20 September 2020).
2. Reagen, B.; Whatmough, P.; Adolf, R.; Rama, S.; Lee, H.; Lee, S.K.; Hernández-Lobato, J.M.; Wei, G.Y.; Brooks, D. Minerva: Enabling low-power, highly-accurate deep neural network accelerators. In *ACM SIGARCH Computer Architecture News*; IEEE Press: Piscataway, NJ, USA, 2016; Volume 44, pp. 267–278.
3. Chen, Y.H.; Emer, J.; Sze, V. Using dataflow to optimize energy efficiency of deep neural network accelerators. *IEEE Micro* **2017**, *37*, 12–21. [CrossRef]
4. Kim, S.; Howe, P.; Moreau, T.; Alaghi, A.; Ceze, L.; Sathe, V.S. Energy-Efficient Neural Network Acceleration in the Presence of Bit-Level Memory Errors. *IEEE Trans. Circuits Syst. I Regul. Pap.* **2018**, *65*, 4285–4298. [CrossRef]
5. Gokhale, V.; Zaidy, A.; Chang, A.X.M.; Culurciello, E. Snowflake: An efficient hardware accelerator for convolutional neural networks. In Proceedings of the 2017 IEEE International Symposium on Circuits and Systems (ISCAS), Baltimore, MD, USA, 28–31 May 2017; pp. 1–4.
6. Sze, V.; Chen, Y.H.; Yang, T.J.; Emer, J.S. Efficient processing of deep neural networks: A tutorial and survey. *Proc. IEEE* **2017**, *105*, 2295–2329. [CrossRef]
7. Andrae, A.S.; Edler, T. On global electricity usage of communication technology: Trends to 2030. *Challenges* **2015**, *6*, 117–157. [CrossRef]
8. Seok, M.; Jeon, D.; Chakrabarti, C.; Blaauw, D.; Sylvester, D. Pipeline Strategy for Improving Optimal Energy Efficiency in Ultra-Low Voltage Design. In Proceedings of the 48th Design Automation Conference, San Diego, CA, USA, 5–10 June 2011; pp. 990–995.
9. Chen, H.; Manzi, D.; Roy, S.; Chakraborty, K. Opportunistic turbo execution in NTC: Exploiting the paradigm shift in performance bottlenecks. In Proceedings of the 2015 52nd ACM/EDAC/IEEE Design Automation Conference (DAC), San Francisco, CA, USA, 8–12 June 2015; pp. 63:1–63:6.
10. Mugisha, D.M.; Chen, H.; Roy, S.; Chakraborty, K. Resilient Cache Design for Mobile Processors in the Near-Threshold Regime. *J. Low Power Electron.* **2015**, *11*, 112–120. [CrossRef]
11. Basu, P.; Chen, H.; Saha, S.; Chakraborty, K.; Roy, S. SwiftGPU: Fostering Energy Efficiency in a Near-Threshold GPU Through a Tactical Performance Boost. In Proceedings of the 2016 53nd ACM/EDAC/IEEE Design Automation Conference (DAC), Austin, TX, USA, 5–9 June 2016.
12. Pinckney, N.; Blaauw, D.; Sylvester, D. Low-power near-threshold design: Techniques to improve energy efficiency energy-efficient near-threshold design has been proposed to increase energy efficiency across a wid. *IEEE Solid State Circuits Mag.* **2015**, *7*, 49–57. [CrossRef]
13. Pinckney, N.R.; Sewell, K.; Dreslinski, R.G.; Fick, D.; Mudge, T.N.; Sylvester, D.; Blaauw, D. Assessing the performance limits of parallelized near-threshold computing. In Proceedings of the 49th Annual Design Automation Conference, San Francisco, CA, USA, 3–7 June 2012; pp. 1147–1152.
14. Dreslinski, R.G.; Wieckowski, M.; Blaauw, D.; Sylvester, D.; Mudge, T.N. Near-Threshold Computing: Reclaiming Moore's Law Through Energy Efficient Integrated Circuits. *Proc. IEEE* **2010**, *98*, 253–266. [CrossRef]
15. Shabanian, T.; Bal, A.; Basu, P.; Chakraborty, K.; Roy, S. ACE-GPU: Tackling Choke Point Induced Performance Bottlenecks in a Near-Threshold Computing GPU. In Proceedings of the International Symposium on Low Power Electronics and Design (ISLPED '18), Bellevue, WA, USA, 23–25 July 2018.

16. Trapani Possignolo, R.; Ebrahimi, E.; Ardestani, E.K.; Sankaranarayanan, A.; Briz, J.L.; Renau, J. GPU NTC Process Variation Compensation With Voltage Stacking. *IEEE Trans. Very Large Scale Integr. Syst.* **2018**, *26*, 1713–1726. [CrossRef]

17. Esmaeilzadeh, H.; Blem, E.; Amant, R.S.; Sankaralingam, K.; Burger, D. Dark silicon and the end of multicore scaling. In Proceedings of the 2011 38th Annual International Symposium on Computer Architecture (ISCA), San Jose, CA, USA, 4–8 June 2011.

18. Silvano, C.; Palermo, G.; Xydis, S.; Stamelakos, I.S. Voltage island management in near threshold manycore architectures to mitigate dark silicon. In Proceedings of the 2014 Design, Automation & Test in Europe Conference & Exhibition (DATE), Dresden, Germany, 24–28 March 2014; pp. 1–6.

19. Chen, Y.H.; Krishna, T.; Emer, J.S.; Sze, V. Eyeriss: An energy-efficient reconfigurable accelerator for deep convolutional neural networks. *IEEE J. Solid State Circuits* **2016**, *52*, 127–138. [CrossRef]

20. Yoo, H.J.; Park, S.; Bong, K.; Shin, D.; Lee, J.; Choi, S. A 1.93 tops/w scalable deep learning/inference processor with tetra-parallel mimd architecture for big data applications. In Proceedings of the IEEE International Solid-State Circuits Conference, San Francisco, CA, USA, 22–26 February 2015; pp. 80–81.

21. Cavigelli, L.; Gschwend, D.; Mayer, C.; Willi, S.; Muheim, B.; Benini, L. Origami: A convolutional network accelerator. In Proceedings of the 25th Edition on Great Lakes Symposium on VLSI, Pittsburgh, PA, USA, 20–22 May 2015; pp. 199–204.

22. Jouppi, N.P.; Young, C.; Patil, N.; Patterson, D.; Agrawal, G.; Bajwa, R.; Bates, S.; Bhatia, S.; Boden, N.; Borchers, A.; et al. In-datacenter performance analysis of a tensor processing unit. In Proceedings of the 2017 ACM/IEEE 44th Annual International Symposium on Computer Architecture (ISCA), Toronto, ON, Canada, 24–28 June 2017; pp. 1–12.

23. Du, Z.; Fasthuber, R.; Chen, T.; Ienne, P.; Li, L.; Luo, T.; Feng, X.; Chen, Y.; Temam, O. ShiDianNao: Shifting vision processing closer to the sensor. In Proceedings of the 2015 ACM/IEEE 42nd Annual International Symposium on Computer Architecture (ISCA), Portland, OR, USA, 13–17 June 2015; pp. 92–104. [CrossRef]

24. Chen, T.; Du, Z.; Sun, N.; Wang, J.; Wu, C.; Chen, Y.; Temam, O. Diannao: A small-footprint high-throughput accelerator for ubiquitous machine-learning. *ACM Sigarch Comput. Archit. News* **2014**, *42*, 269–284. [CrossRef]

25. Zhang, J.; Wang, Z.; Verma, N. A machine-learning classifier implemented in a standard 6T SRAM array. In Proceedings of the 2016 IEEE Symposium on VLSI Circuits (VLSI-Circuits), Honolulu, HI, USA, 15–17 June 2016; pp. 1–2.

26. Shafiee, A.; Nag, A.; Muralimanohar, N.; Balasubramonian, R.; Strachan, J.P.; Hu, M.; Williams, R.S.; Srikumar, V. ISAAC: A convolutional neural network accelerator with in-situ analog arithmetic in crossbars. *ACM Sigarch Comput. Archit. News* **2016**, *44*, 14–26. [CrossRef]

27. Chi, P.; Li, S.; Xu, C.; Zhang, T.; Zhao, J.; Liu, Y.; Wang, Y.; Xie, Y. Prime: A novel processing-in-memory architecture for neural network computation in reram-based main memory. In Proceedings of the 43rd International Symposium on Computer Architecture, Seoul, Korea, 18–22 June 2016.

28. Song, L.; Qian, X.; Li, H.; Chen, Y. Pipelayer: A pipelined reram-based accelerator for deep learning. In Proceedings of the 2017 IEEE International Symposium on High Performance Computer Architecture (HPCA), Austin, TX, USA, 4–8 February 2017; pp. 541–552.

29. Mythic Technology: An Architecture Built from the Ground up for AI. Available online: https://www.mythic-ai.com/technology/ (accessed on 20 September 2020).

30. Graphcore IPU: Designed for Machine Intelligence. Available online: https://www.graphcore.ai/products/ipu (accessed on 20 September 2020).

31. Perceive Ergo: A Complete Solution for High Accuracy, Low Power Intelligence to Consumer Products. Available online: https://perceive.io/product/ (accessed on 20 September 2020).

32. Edge TPU: Google's Purpose-Built ASIC Designed to Run Inference at the Edge. Available online: https://cloud.google.com/edge-tpu (accessed on 20 September 2020).

33. Google Coral Edge TPU Explained in Depth. Available online: https://qengineering.eu/google-corals-tpu-explained.html (accessed on 20 September 2020).

34. Jiao, X.; Luo, M.; Lin, J.H.; Gupta, R.K. An assessment of vulnerability of hardware neural networks to dynamic voltage and temperature variations. In Proceedings of the 36th International Conference on Computer-Aided Design, Irvine, CA, USA, 13–16 November 2017; pp. 945–950.

35. Zhang, J.; Rangineni, K.; Ghodsi, Z.; Garg, S. ThUnderVolt: Enabling Aggressive Voltage Underscaling and Timing Error Resilience for Energy Efficient Deep Neural Network Accelerators. *arXiv* **2018**, arXiv:1802.03806.

36. Ernst, D.; Kim, N.S.; Das, S.; Pant, S.; Rao, R.R.; Pham, T.; Ziesler, C.H.; Blaauw, D.; Austin, T.M.; Flautner, K.; et al. Razor: A Low-Power Pipeline Based on Circuit-Level Timing Speculation. In Proceedings of the 36th Annual IEEE/ACM International Symposium on Microarchitecture (MICRO-36), San Diego, CA, USA, 5 December 2003; pp. 7–18.

37. Das, S.; Tokunaga, C.; Pant, S.; Ma, W.H.; Kalaiselvan, S.; Lai, K.; Bull, D.; Blaauw, D. RazorII: In Situ Error Detection and Correction for PVT and SER Tolerance. *J. Solid State Circ.* **2009**, *44*, 32–48. [CrossRef]

38. Fojtik, M.; Fick, D.; Kim, Y.; Pinckney, N.; Harris, D.; Blaauw, D.; Sylvester, D. Bubble Razor: An architecture-independent approach to timing-error detection and correction. In Proceedings of the 2012 IEEE International Solid-State Circuits Conference, San Francisco, California, USA, 19–23 February 2012; pp. 488–490.

39. Albericio, J.; Judd, P.; Hetherington, T.; Aamodt, T.; Jerger, N.E.; Moshovos, A. Cnvlutin: Ineffectual -Neuron-Free Deep Neural Network Computing. In Proceedings of the 2016 ACM/IEEE 43rd Annual International Symposium on Computer Architecture (ISCA), Seoul, Korea, 18–22 June 2016; pp. 1–13. [CrossRef]

40. Pandey, P.; Basu, P.; Chakraborty, K.; Roy, S. GreenTPU: Improving Timing Error Resilience of a Near-Threshold Tensor Processing Unit. In Proceedings of the 2019 56th ACM/IEEE Design Automation Conference (DAC), Las Vegas, NV, USA, 2–6 June 2019; pp. 173:1–173:6.

41. Kim, S.; Howe, P.; Moreau, T.; Alaghi, A.; Ceze, L.; Sathe, V. MATIC: Learning around errors for efficient low-voltage neural network accelerators. In Proceedings of the 2018 Design, Automation & Test in Europe Conference & Exhibition (DATE), Dresden, Germany, 19–23 March 2018; pp. 1–6.

42. Zhang, J.J.; Garg, S. FATE: fast and accurate timing error prediction framework for low power DNN accelerator design. In Proceedings of the 2018 IEEE/ACM International Conference on Computer-Aided Design (ICCAD); San Diego, CA, USA, 5–8 November 2018; pp. 1–8.

43. Reagen, B.; Gupta, U.; Pentecost, L.; Whatmough, P.; Lee, S.K.; Mulholland, N.; Brooks, D.; Wei, G.Y. Ares: A framework for quantifying the resilience of deep neural networks. In Proceedings of the 2018 55th ACM/ESDA/IEEE Design Automation Conference (DAC), San Francisco, CA, USA, 24–28 June 2018; pp. 1–6.

44. Zhang, J.; Ghodsi, Z.; Rangineni, K.; Garg, S. Enabling Timing Error Resilience for Low-Power Systolic-Array Based Deep Learning Accelerators. *IEEE Des. Test* **2019**, *37*, 93–102. [CrossRef]

45. Gundi, N.D.; Shabanian, T.; Basu, P.; Pandey, P.; Roy, S.; Chakraborty, K.; Zhang, Z. EFFORT: Enhancing Energy Efficiency and Error Resilience of a Near-Threshold Tensor Processing Unit. In Proceedings of the 2020 25th Asia and South Pacific Design Automation Conference (ASP-DAC), Beijing, China, 13–16 January 2020; pp. 241–246.

46. Zhao, W.; Cao, Y. New Generation of Predictive Technology Model for sub-45nm Early Design Exploration. *Electron. Devices* **2006**, *53*, 2816–2823. [CrossRef]

47. Karpuzcu, U.R.; Kolluru, K.B.; Kim, N.S.; Torrellas, J. VARIUS-NTV: A microarchitectural model to capture the increased sensitivity of manycores to process variations at near-threshold voltages. In Proceedings of the IEEE/IFIP International Conference on Dependable Systems and Networks (DSN 2012), Boston, MA, USA, 25–28 June 2012; pp. 1–11.

48. Keras. Available online: https://keras.io (accessed on 20 September 2020).

49. LeCun, Y.; Cortes, C. MNIST Handwritten Digit Database. Available online: http://yann.lecun.com/exdb/mnist/ (accessed on 20 September 2020).

50. Reuters-21578 Dataset. Available online: http://kdd.ics.uci.edu/databases/reuters21578/reuters21578.html (accessed on 20 September 2020).

51. Krizhevsky, A. Learning Multiple Layers of Features from Tiny Images. Available online: http://citeseerx.ist.psu.edu/viewdoc/download?doi=10.1.1.222.9220&rep=rep1&type=pdf (accessed on 20 September 2020).

52. Maas, A.L.; Daly, R.E.; Pham, P.T.; Huang, D.; Ng, A.Y.; Potts, C. Learning Word Vectors for Sentiment Analysis. Available online: https://www.aclweb.org/anthology/P11-1015.pdf (accessed on 20 September 2020).

53. Netzer, Y.; Wang, T.; Coates, A.; Bissacco, A.; Wu, B.; Ng, A.Y. Reading Digits in Natural Images with Unsupervised Feature Learning. In Proceedings of the NIPS Workshop on Deep Learning and Unsupervised Feature Learning, Granada, Spain, 12–17 December 2011.

54. Stallkamp, J.; Schlipsing, M.; Salmen, J.; Igel, C. Man vs. computer: Benchmarking machine learning algorithms for traffic sign recognition. *Neural Netw.* **2012**, *32*, 323–332. [CrossRef]

55. Xiao, H.; Rasul, K.; Vollgraf, R. Fashion-MNIST: A Novel Image Dataset for Benchmarking Machine Learning Algorithms. *arXiv* **2017**, arXiv:1708.07747.

56. Free Spoken Digit Dataset (FSDD). Available online: https://github.com/Jakobovski/free-spoken-digit-dataset (accessed on 20 September 2020).

57. Kim, J.S.; Yang, J.S. DRIS-3: Deep Neural Network Reliability Improvement Scheme in 3D Die-Stacked Memory based on Fault Analysis. In Proceedings of the 2019 56th ACM/IEEE Design Automation Conference (DAC), Las Vegas, NV, USA, 2–6 June 2019; pp. 1–6.

58. Chandramoorthy, N.; Swaminathan, K.; Cochet, M.; Paidimarri, A.; Eldridge, S.; Joshi, R.; Ziegler, M.; Buyuktosunoglu, A.; Bose, P. Resilient Low Voltage Accelerators for High Energy Efficiency. In Proceedings of the 2019 IEEE International Symposium on High Performance Computer Architecture (HPCA), Washington, DC, USA, 16–20 February 2019; pp. 147–158. [CrossRef]

59. Yin, S.; Tang, S.; Lin, X.; Ouyang, P.; Tu, F.; Zhao, J.; Xu, C.; Li, S.; Xie, Y.; Wei, S.; et al. Parana: A parallel neural architecture considering thermal problem of 3d stacked memory. *IEEE Trans. Parallel Distrib. Syst.* **2018**, *30*, 146–160. [CrossRef]

60. Nguyen, D.T.; Kim, H.; Lee, H.J.; Chang, I.J. An approximate memory architecture for a reduction of refresh power consumption in deep learning applications. In Proceedings of the 2018 IEEE International Symposium on Circuits and Systems (ISCAS), Florence, Italy, 27–30 May 2018; pp. 1–5.

61. Salami, B.; Unsal, O.S.; Kestelman, A.C. On the resilience of rtl nn accelerators: Fault characterization and mitigation. In Proceedings of the 2018 30th International Symposium on Computer Architecture and High Performance Computing (SBAC-PAD), Lyon, France, 24–27 September 2018; pp. 322–329.

62. Li, G.; Hari, S.K.S.; Sullivan, M.; Tsai, T.; Pattabiraman, K.; Emer, J.; Keckler, S.W. Understanding error propagation in deep learning neural network (DNN) accelerators and applications. In Proceedings of the International Conference for High Performance Computing, Networking, Storage and Analysis, Denver, CO, USA, 12–17 November 2017; pp. 1–12.

63. Libano, F.; Wilson, B.; Anderson, J.; Wirthlin, M.; Cazzaniga, C.; Frost, C.; Rech, P. Selective hardening for neural networks in fpgas. *IEEE Trans. Nucl. Sci.* **2018**, *66*, 216–222. [CrossRef]

64. Choi, W.; Shin, D.; Park, J.; Ghosh, S. Sensitivity Based Error Resilient Techniques for Energy Efficient Deep Neural Network Accelerators. In Proceedings of the 56th Annual Design Automation Conference 2019, DAC '19, Las Vegas, NV, USA, 2–6 June 2019; pp. 204:1–204:6. [CrossRef]

65. Zhang, J.J.; Gu, T.; Basu, K.; Garg, S. Analyzing and mitigating the impact of permanent faults on a systolic array based neural network accelerator. In Proceedings of the 2018 IEEE 36th VLSI Test Symposium (VTS), San Francisco, CA, USA, 22–25 April 2018; pp. 1–6. [CrossRef]

66. Eshraghian, J.K.; Kang, S.M.; Baek, S.; Orchard, G.; Iu, H.H.C.; Lei, W. Analog weights in ReRAM DNN accelerators. In Proceedings of the 2019 IEEE International Conference on Artificial Intelligence Circuits and Systems (AICAS), Hsinchu, Taiwan, 18–20 March 2019; pp. 267–271.

67. Ghodrati, S.; Sharma, H.; Kinzer, S.; Yazdanbakhsh, A.; Samadi, K.; Kim, N.S.; Burger, D.; Esmaeilzadeh, H. Mixed-Signal Charge-Domain Acceleration of Deep Neural networks through Interleaved Bit-Partitioned Arithmetic. *arXiv* **2019**, arXiv:1906.11915.

68. Mackin, C.; Tsai, H.; Ambrogio, S.; Narayanan, P.; Chen, A.; Burr, G.W. Weight Programming in DNN Analog Hardware Accelerators in the Presence of NVM Variability. *Adv. Electron. Mater.* **2019**, *5*, 1900026. [CrossRef]

69. Shikata, A.; Sekimoto, R.; Kuroda, T.; Ishikuro, H. A 0.5 V 1.1 MS/sec 6.3 fJ/conversion-step SAR-ADC with tri-level comparator in 40 nm CMOS. *IEEE J. Solid-State Circuits* **2012**, *47*, 1022–1030. [CrossRef]

70. Lin, K.T.; Cheng, Y.W.; Tang, K.T. A 0.5 V 1.28-MS/s 4.68-fJ/conversion-step SAR ADC with energy-efficient DAC and trilevel switching scheme. *IEEE Trans. Very Large Scale Integr. (VLSI) Syst.* **2016**, *24*, 1441–1449. [CrossRef]

71. Liu, C.C.; Chang, S.J.; Huang, G.Y.; Lin, Y.Z. A 10-bit 50-MS/s SAR ADC with a monotonic capacitor switching procedure. *IEEE J. Solid State Circuits* **2010**, *45*, 731–740. [CrossRef]
72. Song, J.; Jun, J.; Kim, C. A 0.5 V 10-bit 3 MS/s SAR ADC With Adaptive-Reset Switching Scheme and Near-Threshold Voltage-Optimized Design Technique. *IEEE Trans. Circuits Syst. II Express Briefs* **2020**, *67*, 1184–1188. [CrossRef]
73. Hong, H.C.; Lin, L.Y.; Chiu, Y. Design of a 0.20–0.25-V, sub-nW, rail-to-rail, 10-bit SAR ADC for self-sustainable IoT applications. *IEEE Trans. Circuits Syst. I Regul. Pap.* **2019**, *66*, 1840–1852. [CrossRef]
74. Sheu, S.S.; Chang, M.F.; Lin, K.F.; Wu, C.W.; Chen, Y.S.; Chiu, P.F.; Kuo, C.C.; Yang, Y.S.; Chiang, P.C.; Lin, W.P.; et al. A 4Mb embedded SLC resistive-RAM macro with 7.2 ns read-write random-access time and 160 ns MLC-access capability. In Proceedings of the 2011 IEEE International Solid-State Circuits Conference, San Francisco, CA, USA, 20–24 February 2011; pp. 200–202.
75. Niu, D.; Xu, C.; Muralimanohar, N.; Jouppi, N.P.; Xie, Y. Design trade-offs for high density cross-point resistive memory. In Proceedings of the 2012 ACM/IEEE International Symposium on LOW Power Electronics and Design, Redondo Beach, CA, USA, 30 July–1 August 2012; pp. 209–214.
76. Chang, M.F.; Wu, J.J.; Chien, T.F.; Liu, Y.C.; Yang, T.C.; Shen, W.C.; King, Y.C.; Lin, C.J.; Lin, K.F.; Chih, Y.D.; et al. 19.4 embedded 1Mb ReRAM in 28 nm CMOS with 0.27-to-1V read using swing-sample-and-couple sense amplifier and self-boost-write-termination scheme. In Proceedings of the 2014 IEEE International Solid-State Circuits Conference Digest of Technical Papers (ISSCC), San Francisco, CA, USA, 9–13 February 2014.
77. Chang, M.F.; Wu, C.W.; Hung, J.Y.; King, Y.C.; Lin, C.J.; Ho, M.S.; Kuo, C.C.; Sheu, S.S. A low-power subthreshold-to-superthreshold level-shifter for sub-0.5 V embedded resistive RAM (ReRAM) macro in ultra low-voltage chips. In Proceedings of the 2014 IEEE Asia Pacific Conference on Circuits and Systems (APCCAS), Ishigaki, Japan, 17–20 November 2014.
78. Chang, M.F.; Wu, C.W.; Kuo, C.C.; Shen, S.J.; Lin, K.F.; Yang, S.M.; King, Y.C.; Lin, C.J.; Chih, Y.D. A 0.5 V 4 Mb logic-process compatible embedded resistive RAM (ReRAM) in 65nm CMOS using low-voltage current-mode sensing scheme with 45ns random read time. In Proceedings of the 2012 IEEE International Solid-State Circuits Conference, San Francisco, CA, USA, 19–23 February 2012.
79. Ishii, T.; Ning, S.; Tanaka, M.; Tsurumi, K.; Takeuchi, K. Adaptive comparator bias-current control of 0.6 V input boost converter for ReRAM program voltages in low power embedded applications. *IEEE J. Solid State Circuits* **2016**, *51*, 2389–2397. [CrossRef]

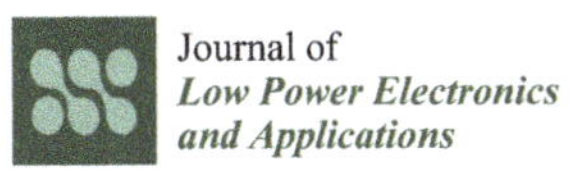

Journal of
Low Power Electronics and Applications

Article

Idleness-Aware Dynamic Power Mode Selection on the i.MX 7ULP IoT Edge Processor

Alfio Di Mauro [1],*, Hamed Fatemi [2], Jose Pineda de Gyvez [3] and Luca Benini [1]

[1]   Integrated Systems Laboratory, ETHZ, 8092 Zurich, Switzerland; lbenini@iis.ee.ethz.ch
[2]   NXP Semiconductor, San Jose, CA 95134, USA; hamed.fatemi@nxp.com
[3]   NXP Semiconductor, 5656 Eindhoven, The Netherlands; jose.pineda.de.gyvez@nxp.com
*   Correspondence: adimauro@iis.ee.ethz.ch

Received: 15 April 2020; Accepted: 20 May 2020; Published: 5 June 2020

Abstract: Power management is a crucial concern in micro-controller platforms for the Internet of Things (IoT) edge. Many applications present a variable and difficult to predict workload profile, usually driven by external inputs. The dynamic tuning of power consumption to the application requirements is indeed a viable approach to save energy. In this paper, we propose the implementation of a power management strategy for a novel low-cost low-power heterogeneous dual-core SoC for IoT edge fabricated in 28 nm FD-SOI technology. Ss with more complex power management policies implemented on high-end application processors, we propose a power management strategy where the power mode is dynamically selected to ensure user-specified target idleness. We demonstrate that the dynamic power mode selection introduced by our power manager allows achieving more than 43% power consumption reduction with respect to static worst-case power mode selection, without any significant penalty in the performance of a running application.

Keywords: edge devices; power management; energy efficiency

---

1. Introduction

Efficient energy management is very challenging in Internet of Things (IoT) edge devices [1]. On one hand there is the increasing demand of more near-sensor computing capabilities, on the other hand, strict constraints have to be set on the power consumption to maximize the lifetime of an IoT node, which is in many cases battery-supplied.

Researchers have responded to this challenge by proposing new SoC architectures, e.g., parallel-ultra-low-power multi-core computing platforms, and power management strategies [2]. Similar approaches are proposed also by industry, e.g., by adopting heterogeneous multi-core architectures where a low-power control core (e.g., Arm® Cortex®-M4 core (Arm and Cortex are registered trademarks of Arm Limited (or its subsidiaries) in the US and/or elsewhere)) with modest computing capabilities is coupled with a more capable core (e.g., Arm Cortex-A7 core) for compute-intensive tasks.

In this scenario, well-known techniques such as clock-gating and power-gating are widely used to minimize the power consumption of inactive sub-modules of an SoC. However, restoring the functionality of SoC subsystems that are clock-gated or power-gated during energy saving states could require a non-negligible amount of time and energy. Additionally, in many event-driven applications, requiring fast time response or continuous event processing, duty-cycling computation phases, or even occasionally entering a sleep state can be unfeasible.

Shut-down-based power management can be complemented by more advanced energy saving techniques such as Dynamic Voltage and Frequency Scaling (DVFS). DVFS is the process of adapting, at run-time, the frequency and the supply voltage of a digital circuit; typically, this is obtained in a closed loop regulation, using as a feedback parameters such as core workload or desired core idleness. The dynamic power consumption of digital circuits has linear dependency from the frequency, and it is in a quadratic relationship with the supply voltage. Therefore, such a technique is very effective to significantly reduce the energy consumption of digital circuits.

Unfortunately, the hardware and software infrastructure for DVFS available in heterogeneous edge devices is not yet as mature as the one integrated into many high-end and mobile multi-core application processors. Such development is not trivial since, in sharp contrast with high-end multi-cores, dual-core MCUs have most of the times heterogeneous ISAs, non-uniform memory hierarchy and neither cache coherency nor shared virtual memory support.

The NXP i.MX 7ULP is a low-power low-cost Arm Cortex-M4/Cortex-A7-based SoC which belongs to this device category. The chip has been fabricated in 28 nm Fully Depleted Silicon On Insulator (FD-SOI) technology, and it features an advanced power management infrastructure that enables dynamic power mode adaptation. The main contribution of this paper is the development and qualification of the first (to our knowledge) DVFS-based power management software infrastructure for this exemplary heterogeneous dual-core MCUs. Our power manager achieves up to 45% of power consumption reduction during the active state of the SoC, without any penalty on the application execution performance.

The remainder of this paper is organized as follows. Section 2 gives an overview of the related work. Section 3 describes the i.MX 7ULP SoC. Section 4 describes the power manager. In Section 5 we describe the experimental setup used to characterize the power manager. Section 6 presents the results in terms of power consumption reduction, while Section 7 provides concluding remarks.

2. Related Work

Among several techniques to increase the energy efficiency, Dynamic Voltage and Frequency Scaling (DVFS) represents a well established, effective [3] and low-complexity strategy. Despite technological scaling, which reduces the margins for supply voltage regulation, DVFS still represents a viable energy saving techniques for modern platforms operating near threshold [4]. Those platforms reach high energy efficiencies thanks to a combination of voltage scaling and adaptive body biasing to avoid performance degradation [5,6]. In [7,8], authors present two approaches where DVFS is controlled by dedicated hardware units on embedded microprocessors. Similarly, in [9] DVFS is applied on high-end CPUs for cloud infrastructures.

More sophisticated techniques rely on gathering information on the status of the system to guarantee a desired quality of service. In [10], authors exploit a combination of DVFS and workload profiling in a closed loop configuration.In [11,12], to optimize the efficiency of Linux-based systems, DVFS is operated in combination with advanced power-aware scheduling algorithms that exploit a combination of system-level statistics. Finally, in [13] machine learning algorithms are used to gather higher level information from the applications running on the core [13].

In this paper, we propose a power management unit implementation that takes inspiration from the power management policies surveyed above and adopted on many Linux-capable application processors. The proposed power manager can perform DVFS on a new low-cost low-power ARM A7-M4 SoC architecture for IoT edge (e.g., the NXP i.MX 7ULP SoC). Compared to the available power management implementations [14], our approach targets lower-end devices where techniques such as DVFS are typically not applied because of the lack of hardware/software support. As with what happens in high-end CPUs based systems, the proposed power management gathers system level information, e.g., the idleness of the system, and exploit this information to optimize the use of available energy. The regulation of the power mode is operated by a software implemented power manager running in a low-power real-time domain of the SoC. Unlike application-specific DVFS techniques used in special-purpose accelerators, our approach relies on minimal knowledge of the system, since it exploits very high-level information such as the CPU idle time, making it suitable to be ported on different platforms with similar hardware features.

3. The i.MX 7ULP Platform

i.MX 7ULP is a new low-cost low-power Arm Cortex-M4/Cortex-A7-based SoC architecture for IoT edge. The chip has been developed by NXP and fabricated in 28 nm FD-SOI technology. The architecture of the microprocessor is divided into two processing domains hosted by two separate power domains. (i) The application domain is built around an Arm Cortex-A7 core. This processing domain can boot Linux-based embedded operating systems, offering high flexibility for application development. The main target of this processing domain are general purpose and computationally demanding workloads. (ii) The real-time domain is built around an Arm Cortex-M4 core. This domain targets low-power low-complexity sections of an application that presents strict requirements in terms of predictable tasks execution time, e.g., timing-critical tasks such as sensor sampling and IO event handling.

The SoC features flexible hardware support for power management. Each of the two power domains has a dedicated register set that can be used to store pre-defined parameters such as supply voltage, clock frequency divider and body-biasing. The switching between pre-configured power modes is governed by a dedicated System Mode Controller (SMC).

4. Power Manager

The goal of the power manager proposed in this paper is to minimize the power consumption when the system is in an active state, all the buses are clocked, and cores have complete access to all peripherals. In the rest of the paper, we will refer to this system configuration as *RUN* power state. The power manager operates an opportunistic voltage and frequency scaling on the application power domain to reduce the energy consumption when the average workload of the A7 core is lower than 100%. In this context, a statically selected power mode, tailored for a worst-case workload, is inefficient, and the same task could be executed operating the core at a lower voltage and frequency.

To limit the intrusiveness of the approach, and to have a deterministic time response to workload variations, we implemented the power manager on the real-time application domain. More specifically, the power manager is part of the firmware application running on the M4 core, which serves as a software Power Management Unit (PMU). Figure 1 illustrates the block diagram of the power manager and the main sub-modules.

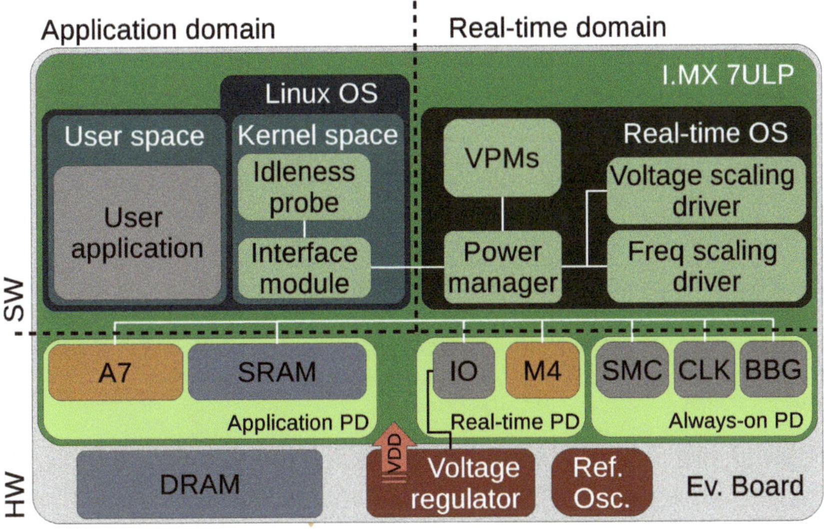

Figure 1. Power manager software architecture block diagram and Hardware/Software partitioning.

To operate the DVFS, we statically selected the RUN power mode. In this power state, the system uses the main PLL to generate a reference clock frequency in the range from 800 MHz to 300 MHz; the forward body-bias can be applied to the transistors, the RAM is powered on, and the system can tolerate DVFS. In this scenario, we introduced multiple *Virtual Power Modes (VPM)*. Each VPM is a set of values that the power manager uses to override configuration values for Voltage, clock Frequency and Body-bias activation of the *RUN* power mode. The number of VPMs can be arbitrarily configured by the user, their configuration values are stored in the M4 data memory.

The criterion on which the VPM of the system is switched is the idleness of the A7 core. This information is obtained directly from the operating system CPU statistics. The power manager compares the measured idleness with a user-specified target idleness, and determines how to adjust the performance of the system to match the target idleness. Algorithm 1 describes the performance adjustment procedure. The idleness observation time can vary in a range from 100 ms to 5 s depending on the application requirements.

Algorithm 1 This pseudo-code procedure illustrates the steps that are performed by the power manager at every iteration of the power mode regulation.

```
 1: while !new power mode request do
 2:    wait for a new request
 3: end while
 4: if target idleness > curr. idleness then
 5:    increase supply voltage
 6:    while !Supply voltage change do
 7:       wait
 8:    end while
 9:    increase clock freq.
10:    while !clock freq. change do
11:       wait
12:    end while
13: else
14:    decrease clock freq.
15:    while !clock freq. change do
16:       wait
17:    end while
18:    decrease supply voltage
19:    while !Supply voltage change do
20:       wait
21:    end while
22: end if
23: power mode transition complete
```

4.1. Real-Time Domain

The kernel of the power manager is part of the firmware running on the real-time domain, and it is constituted by 4 main sub-modules that operate synchronously to perform (i) the idleness acquisition from the application domain, (ii) the target versus current idleness comparison, (iii) the power mode selection and the power mode configuration registers modification, (iv) the voltage and the frequency scaling.

4.2. Application Domain

The power manager uses the idleness of the core in a given time interval preceding the power mode regulation as a feedback signal. This information is made available by the operating system that collects statistics on the use of the core. To retrieve this information and transfer it to the real-time domain, we developed (i) a module to track the idleness and (ii) a Linux-driver module that acts as an interface with the power manager.

5. Experimental Setup

In this section, we describe the experimental setup for the measurements presented in Section 6. In our experiments, we focused on the dynamic reduction of the energy consumption of the application power domain only, since it represents more than 90% of the power consumption of the entire SoC. Operating frequency and voltage of the real-time power domain have been statically selected as the minimum values that allow accomplishing the power management task.

In our power management implementation, we introduced six virtual power modes available for dynamic power mode selection. Table 1 reports the configuration values related to the core supply voltage and clock frequency of every VPM. Supply voltages have been arbitrarily chosen according to a quadratic curve, operating frequencies have been selected as the maximum ones achievable by the core at the related supply voltages. To maximize performance, forward body-bias is active in all VPMs [15]. To verify that the system was able to operate at a given frequency, we executed several benchmarks on the core, selected among those provided by the *stress-ng* benchmark suite [16].

Table 1. Configuration values of Virtual Power Modes (VPM).

Power Mode ID	VPM0	VPM1	VPM2	VPM3	VPM4	VPM5
Voltage [mV]	1100	1000	900	860	835	800
Frequency [MHz]	800	640	490	420	370	300

The power mode regulation period has been chosen as 5 s. This value can be modified to adjust responsiveness of the controller to the idleness variation, the minimum regulation period is approximately of 100 ms; this value is bounded by the communication latency with the external voltage regulator hosted on the i.MX 7ULP evaluation board and does not constitute an intrinsic limitation of the i.MX 7ULP system.

To fully characterize the power manager and evaluate the benefits in terms of power saving we performed the following measurements on the i.MX 7ULP evaluation board:

(i) we fixed the VPM, and we measured its power consumption. We performed the measurements for each of the VPMs reported in Table 1, and, at each VPM, we swept the CPU usage in a range of 5% to 95%. The desired load of the core has been precisely changed by varying the number of tasks launched in a fixed time period of 5 s. The use information of the core has been obtained by accessing the core use and idleness statistic made available by the Linux kernel. In our experiment, we verified that by executing a small number of tasks the core use was below 5%, and we linearly increased the number of task launched on the core within the reference period until the core use reached 95%.

(ii) we evaluated the power consumption at different workload levels when the controller is adjusting the power mode. We analyzed the response of the power manager when the target idleness was in a range of 5% to 95%, and similarly to the previous case, the workload on the core was swept in a range of 5% to 95%.

Section 6 shows the results of the two experiments; the power consumption has been measured at the evaluation board 5 V power supply connector. The measurement conditions are the following:

- Operating temperature T = 25 °C
- Test benchmark for operating frequency validation and power estimation = stress-ng test suite
- Power measurement equipment = Keysight N6705B power analyzer (time resolution of 20 μs)

6. Results

In this section, we present the results in terms of power consumption reduction when the power manager is active.

Figure 2a shows the power consumption of the i.MX 7ULP evaluation board when a VPM is fixed. From the static analysis, we can conclude that the dynamic power mode adaptation allows saving up to 28% of power when the system is in RUN mode, and in absence of core activity.

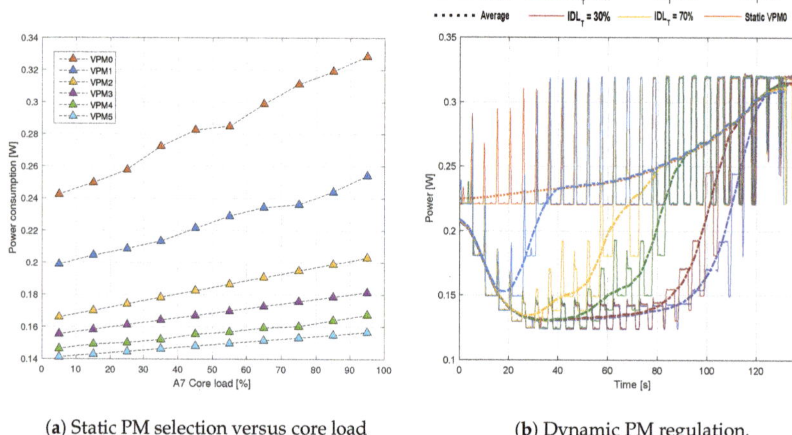

(**a**) Static PM selection versus core load (**b**) Dynamic PM regulation.

Figure 2. Core power consumption.

The power consumption reported in the plot includes the power consumption of other components soldered on the board, e.g., the external regulator. The only contribution excluded from this analysis is the power consumption of the on-board external DRAM chip.

In the second experiment we measured the power consumption of the board when the power manager is regulating the VPM. Figure 2b reports the power consumption when the workload on the core increases linearly from from 5% (leftmost part of the plot) to 95% (rightmost part of the plot). The goal of this experiment is show, at run-time, how the PMU can promptly adapt the VPM to the workload scenario, without reducing the number of tasks completed in a reference time interval, and eventually settling on a desired average idleness level. Figure 2b shows that regardless of the selected target idleness, tasks can be released always with the same timing on the core. During the dynamic VPM adaptation, the actual core use depends on the power mode selected by the controller, which tries to restore the target idleness value. In this operating mode, the time to complete a tasks is not known in advance, because the CPU operates at a variable frequency. Please note that the quality of service, which in this context is represented by meeting the task deadlines is not degraded when the VPM is dynamically changed (i.e., the core computational bandwidth is never saturated). The Figure 2b reports that the average power consumption is always equal or lower than the case where a VPM is statically selected. Always in Figure 2b, it can be observed that the target idleness specified by the user affects the power manager behavior. Table 2 summarizes the peak power consumption reduction for several user-specified values of target idleness.

Table 2. Peak power consumption reduction with respect to static VPM0 selection.

Target Idleness	90%	70%	50%	30%	10%
Peak power reduction	32%	33%	43%	44%	45%

From the results reported in Table 2, it is evident that when the target idleness is too high with respect to the actual core workload, the power manager cannot explore the entire set of available states. On the contrary, in all those cases where the power manager can operate the system at the lowest power modes (Target idleness = 50%, 30%, 10%), it is possible to achieve more than 43% of average power consumption reduction and a peak power consumption reduction of 45%. The power consumption improvement reported in our measurements refers to the case where VPM0 was statically selected, and the core executed the same workload.

In our experiment, we assumed that the number of operation performed at each task execution is constant. In Figure 2b, we observe that every 5 s, a new set of task is released on the core, originating a spike in the power trace caused by the core wake-up. At each experiment run, which was performed with a different target idleness, the core never saturated the computational bandwidth. Therefore, the number of task executed by the core was constant for all the experiments, while the power consumption significantly decreased, resulting in less energy consumed by the core.

7. Conclusions

In this paper, we presented the implementation of a power management unit capable of operating an idleness aware dynamic power mode adaptation. We validated the proposed approach on a novel low-cost low-power heterogeneous MCU for IoT edge fabricated in 28 nm SOI technology. The proposed power manager allows achieving a theoretical power consumption reduction higher than the 28% in absence of core activity. We confirmed the expectations during the operation of the system, reporting more than 43% of average power consumption reduction when the target idleness is selected according to the real idleness of the application, and a peak power consumption reduction of 45%.

Author Contributions: Conceptualization, L.B. and A.D.; methodology, H.F. and J.P.; software, A.D.; validation, A.D.; investigation, A.D.; data curation, A.D.; writing–original draft preparation, A.D.; writing–review and editing, A.D., H.F., J.P. and L.B.; supervision, H.F., J.P. and L.B.; project administration, H.F., J.P. and L.B. All authors have read and agree to the published version of the manuscript.

Funding: Partially supported by NXP.

Conflicts of Interest: The authors declare no conflict of interest.

Abbreviations

The following abbreviations are used in this manuscript:

IoT	Internet of Things
MCU	Micro Controller Unit
SoC	System on Chip
DVFS	Dynamic voltage and Frequency Scaling
ISA	Instruction Set Architecture
FD-SOI	Fully Depleted Silicon On Insulator
ULP	Ultra Low Power
CPU	Central Processing Unit
SMC	System Mode Controller
PMU	Power Management Unit
PLL	Phase Locked Loop
VPM	Virtual Power Mode

References

1. Conti, F.; Rossi, D.; Pullini, A.; Loi, I.; Benini, L. PULP: A Ultra-Low Power Parallel Accelerator for Energy-Efficient and Flexible Embedded Vision. *J. Signal Process. Syst.* **2015**, *84*, 339–354. [CrossRef]
2. Rossi, D.; Pullini, A.; Loi, I.; Gautschi, M.; Gurkaynak, F.K.; Teman, A.; Constantin, J.; Burg, A.; Miro-Panades, I.; Beignè, E.; et al. 193 mops/mw @ 162 mops, 0.32 v to 1.15 v voltage range multi-core accelerator for energy efficient parallel and sequential digital processing. In Proceedings of the 2016 IEEE Symposium in Low-Power and High-Speed Chips (COOL CHIPS XIX), Yokohama, Japan, 20–22 April 2016; pp. 1–3.
3. Qu, G. What is the limit of energy saving by dynamic voltage scaling? In Proceedings of the IEEE/ACM International Conference on Computer Aided Design, San Jose, CA, USA, 4–8 November 2001; pp. 560–563.
4. Rossi, D.; Pullini, A.; Loi, I.; Gautschi, M.; Gürkaynak, F.K.; Teman, A.; Constantin, J.; Burg, A.; Miro-Panades, I.; Beignè, E.; et al. Energy-efficient Near-Threshold Parallel Computing : The PULPv2 Cluster. *IEEE Micro* **2017**, *37*, 20–31. [CrossRef]
5. Di Mauro, A.; Rossi, D.; Pullini, A.; Flatresse, P.; Benini, L. Independent Body-Biasing of P-N Transistors in an 28nm UTBB FD-SOI ULP Near-Threshold Multi-Core Cluster. In Proceedings of the 2018 IEEE SOI-3D-Subthreshold Microelectronics Technology Unified Conference (S3S), Burlingame, CA, USA, 15–18 October 2018; pp. 1–3.

6.   Di Mauro, A.; Rossi, D.; Pullini, A.; Flatresse, P.; Benini, L. Temperature and process-aware performance monitoring and compensation for an ULP multi-core cluster in 28 nm UTBB FD-SOI technology. In Proceedings of the 2017 27th International Symposium on Power and Timing Modeling, Optimization and Simulation (PATMOS), Thessaloniki, Greece, 25–27 September 2017; pp. 1–8.

7.   Choi, K.; Soma, R.; Pedram, M. Fine-grained dynamic voltage and frequency scaling for precise energy and performance tradeoff based on the ratio of off-chip access to on-chip computation times. *IEEE Trans. Comput. Aided Des. Integr. Circuits Syst.* **2005**, *24*, 18–28.

8.   Seki, T.; Akui, S.; Seno, K.; Nakai, M.; Meguro, T.; Kondo, T.; Hashiguchi, A.; Kawahara, H.; Kumano, K.; Shimura, M. Dynamic voltage and frequency management for a low-power embedded microprocessor. *IEICE Trans. Electron.* **2005**, *88*, 520–527. [CrossRef]

9.   Krishnaswamy, V.; Brooks, J.; Konstadinidis, G.; McAllister, C.; Pham, H.; Turullols, S.; Shin, J.L.; YangGong, Y.; Zhang, H. Fine-Grained Adaptive Power Management of the SPARC M7 processor. In Proceedings of the 2015 IEEE International Solid-State Circuits Conference (ISSCC), San Francisco, CA, USA, 22–26 February 2015; pp. 74–76.

10.   Cheng, W.H.; Baas, B.M. Dynamic voltage and frequency scaling circuits with two supply voltages. In Proceedings of the IEEE International Symposium on Circuits and Systems, Seattle, WA, USA, 18–21 May 2008; pp. 1236–1239.

11.   Scordino, C.; Lipari, G. A resource reservation algorithm for power-aware scheduling of periodic and aperiodic real-time tasks. *IEEE Trans. Comput.* **2006**, *55*, 1509–1522. [CrossRef]

12.   Spiliopoulos, V.; Kaxiras, S.; Keramidas, G. Green governors: A framework for continuously adaptive dvfs. In Proceedings of the 2011 International Green Computing Conference and Workshops, Orlando, FL, USA, 25–28 July 2011; pp. 1–8.

13.   Juan, D.C.; Garg, S.; Park, J.; Marculescu, D. Learning the optimal operating point for many-core systems with extended range voltage/frequency scaling. In Proceedings of the Ninth IEEE/ACM/IFIP International Conference on Hardware/Software Codesign and System Synthesis (CODES+ISSS '13), Montreal, QC, Canada, 29 September–4 October 2013; IEEE Press: Piscataway, NJ, USA, 2013; pp. 8:1–8:10.

14.   Pallipadi, V.; Starikovskiy, A. The Ondemand Governor. *Linux Symp.* **2006**, *2*, 223–238.

15.   Rossi, D.; Pullini, A.; Loi, I.; Gautschi, M.; Gürkaynak, F.K.; Bartolini, A.; Flatresse, P.; Benini, L. A 60 GOPS/W, -1.8 v to 0.9 v body bias ULP cluster in 28 nm UTBB FD-SOI technology. *Solid-State Electron.* **2016**, *117*, 170–184. [CrossRef]

16.   King, C.I. Stress-ng. Available online: https://github.com/ColinIanKing/stress-ng (accessed on 10 August 2018).

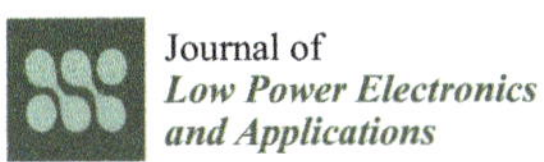

Journal of
*Low Power Electronics
and Applications*

Article

Low-Power Embedded System for Gait Classification Using Neural Networks

Francisco Luna-Perejón *, Manuel Domínguez-Morales, Daniel Gutiérrez-Galán
and Antón Civit-Balcells

Architecture and Computer Technology Department, Universidad de Sevilla, E.T.S Ingeniería Informática,
Reina Mercedes Avenue, 41012 Seville, Spain; mjdominguez@us.es (M.D.-M.); dgutierrez5@us.es (D.G.-G.);
civit@us.es (A.C.-B.)
* Correspondence: fralunper@atc.us.es

Received: 19 March 2020; Accepted: 17 April 2020; Published: 1 May 2020

Abstract: Abnormal foot postures can be measured during the march by plantar pressures in both dynamic and static conditions. These detections may prevent possible injuries to the lower limbs like fractures, ankle sprain or plantar fasciitis. This information can be obtained by an embedded instrumented insole with pressure sensors and a low-power microcontroller. However, these sensors are placed in sparse locations inside the insole, so it is not easy to correlate manually its values with the gait type; that is why a machine learning system is needed. In this work, we analyse the feasibility of integrating a machine learning classifier inside a low-power embedded system in order to obtain information from the user's gait in real-time and prevent future injuries. Moreover, we analyse the execution times, the power consumption and the model effectiveness. The machine learning classifier is trained using an acquired dataset of 3000+ steps from 6 different users. Results prove that this system provides an accuracy over 99% and the power consumption tests obtains a battery autonomy over 25 days.

Keywords: machine learning; neural networks; gait analysis; embedded system

---

1. Introduction

Foot and ankle pain are very common in the population. Studies indicates that around 24% of people aged over 45 years report frequent foot pain [1]. Moreover, other studies indicate that more than 70% of the population over 65 years old present chronic foot pain [2].

It has also been demonstrated that abnormal foot postures and gait are associated with foot pain [3] as well as with lower limb injuries and pathologies [4]. Additionally, problems and disabilities associated with abnormal gait and foot posture include fractures, ankle sprain, pimple pain or plantar fasciitis, among others [5].

These previously named abnormalities due to bad foot postures have recently been related in several experiments to the pressure received at the base of the foot [4,6]. Therefore, a professional specialized in foot problems can perform a walking study of the patient's footprint in order to detect these problems, prevent the injuries occasioned by prescribing insoles and/or indicate physical exercises to correct them.

For that reason, it is very important to characterize the static foot posture and the foot function with a gait analysis. In that concern, there are available various methods in the literature [7].

The classic gait study consists of walking in a straight line through a sensorized surface that emulates a several meter long path—the surface measures and records the pressure obtained for each step during the gait for posterior analysis. The main problem using this mechanism is the psychological component—the patient knows that he/she is being observed and walks, without any intention, in a different way (better or worse, it depends on the patient's mood). So, because of that, in many cases the recorded information does not correspond to the usual patient's way of walking [8].

To avoid the problems found in the classical methods (the psychological component mainly), many recent studies try to embed the sensorized surface in the patient's shoes [9–16].

These developments are mainly focused on designing an instrumented insole that includes pressure sensors, and demonstrate that these devices may have multiple applications in several fields such as in orthopaedic, orthoprosthetic, footwear designing, prostheses, pathology, or even in sports medicine, for the study of the most appropriate footwear in each athletic modality.

As detailed before, the use of instrumented insoles improves the data-recollection process during the gait while the patient is doing his daily-living activities (with freedom of movement and without space limitation). Nevertheless, to achieve good results collecting useful data, these insoles should have a good battery life; otherwise, data will be lost, and the gait analysis study will not be complete.

Additionally, the works developed until now use the footwear insole only to collect data and send it to a processing system like a smartphone or a computer. Due to that, the data is transmitted using a wireless connection in a continuous way and, therefore, the battery life is reduced significantly. Theoretically, if the information is processed locally inside the embedded system, the battery life increases because of the absence of data transmissions—works like that in References [17–20] demonstrate the battery-life improvement.

Recently, we developed an instrumented insole able to receive the pressure information obtained during the gait and send it to a computer via Bluetooth. Running in the computer, a local neural-network system classified the gait type as pronator, supinator or neutral and store that information [21]. Although there is no consensus on the terminology, we will use the common terms "pronation" to indicate when the foot undergoes greater lowering of the medial longitudinal arch and more medial distribution of plantar loading during gait and "supination" when the foot undergoes greater elevation of the medial longitudinal arch and more lateral distribution of plantar loading during gait [3] (see Figure 1).

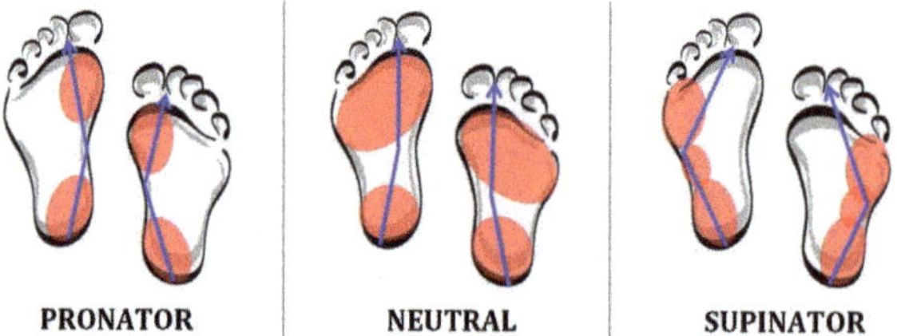

Figure 1. Gait type.

The main goal of that work was to study the feasibility of the proposed gait-type classification, without taking into account any battery-life restrictions. Although we demonstrated that our classification accuracy was better than that obtained in other projects, our work shared the problems related to short battery life.

So, the aim of this work is to the reduce the power-consumption requirements of the instrumented insole by implementing the neural-network classifier into the microcontroller attached to the instrumented insole. To do that, several neural-networks architectures have been trained and tested with Tensorflow and Keras, using a database of 3000+ steps. We evaluate the effectiveness of the classifier in terms of the accuracy, among other metrics. After that, this architecture is compiled and integrated in the embedded system using STM32Cube.AI (artificial intelligence plugin used in STM32CubeIDE software for STMicroelectronics microcontrollers) in order to check the correct behaviour when running on the microcontroller, as well as to assess the power-consumption reduction when classifying with the low-power microcontroller.

The rest of the paper is divided in the following way—first, the acquisition and evaluation processes are described in the Materials and Methods section, presenting the used embedded system , the collected database, and evaluated the neural-networks architectures. Next, the results obtained after the training process with the different neural-networks architectures in Keras, the classification from the neural network deployed into the embedded system and the power-consumption study are detailed and explained in the Results and Discussion section. Finally, conclusions are presented.

2. Materials and Methods

As detailed in the previous section, the system used in this work requires an instrumented insole composed of a set of force sensitive resistors (FSRs) and a microcontroller for the acquisition step and for the final implementation.

Moreover, the main feature that makes the system to reduce drastically the power-consumption requirements consists in the implementation of the machine learning classifier in the microcontroller, avoiding the continuous data transmissions.

All these components and tools will be detailed in depth in this section, as well as the process followed from data acquisition to the final implementation.

2.1. Data Acquisition

The data registered for the gait analysis consists of a set of pressure measures obtained through a footwear insole connected to an embedded device. After an in depth study of the walking process and the more adequate distribution of the sensors tested on previous works [21,22], seven FSRs are disposed in different parts of the foot. For each footstep, sensors samples at 50 hertz frequency since the first contact of the foot with the ground until the moment the foot is lifted. Thus, the information stored refers to the medium pressure received by each sensor during each step. These values are normalized after each step ends using the sensor's value with the highest pressure received as 100% pressure and modifying the other sensors' values to a percentage value relative to that sensor.

Next, both the footwear insole and the dataset obtained after the acquisition phase are detailed.

2.1.1. Footwear Insole

The hardware system used for the acquisition is based on a low-power consumption microcontroller, seven force sensitive resistors (FSRs) and a low-energy Bluetooth module. The selected microcontroller was a STMicroelectronics MCU used for the acquisition and testing phases (model STM32L476RG, operating at a frequency up to 80 MHz, with 1 Mbyte of flash memory and 128 Kbytes of SRAM), with features that allow real-time capabilities, digital signal processing and low-power operation. The FSRs were connected to the analog inputs of the microcontroller using a voltage divider with a 10 KΩ resistor. These sensors provide information about the maximum pressure point and the load forces. Finally, a HM-10 BLE (Bluetooth Low-Energy) module was connected to the microcontroller as a wireless communication port to send the information to the computer (see Figure 2).

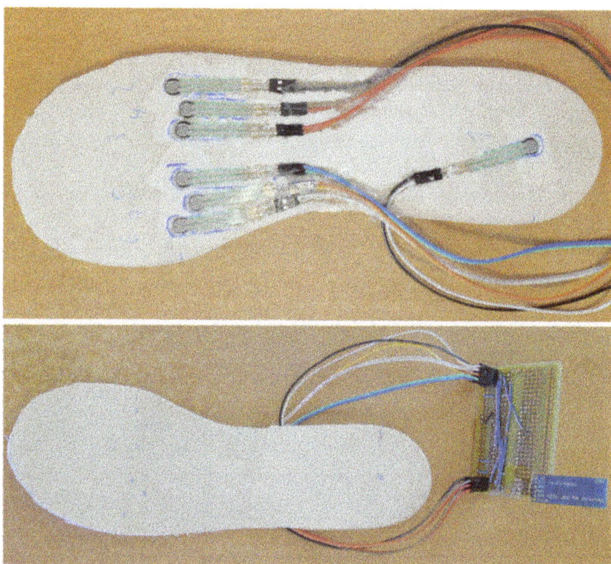

Figure 2. (Left) force sensitive resistors (FSRs) location and microcontroller with wireless module.

The location of the sensors in the footwear insole was established based on the anatomy of the foot, the types of footprints to classify (shown in the previous section) and the tests performed in previous works [22]. Results showed that the metatarsus area gives more information about the footprint types than the other areas, so six sensors were placed in this foot region while one last sensor was placed in the heel to determine the moments of contact with the ground and foot lifted (see Figure 2).

In order to recover the pressure measures for each footstep and obtain an appropriate dataset, the microcontroller implements a FreeRTOS https://www.freertos.org/ based firmware to manage the sensors reading and the communication without information loss. FreeRTOS allows us to manage the implemented functions using OS functionalities, such as semaphores, queues and tasks. The sent data was adequately collected by a computer application. In order to understand the data acquisition process, a graphic diagram is shown in Figure 3-up. In this diagram, the other two phases of this work are also shown. They will be explained later in this paper.

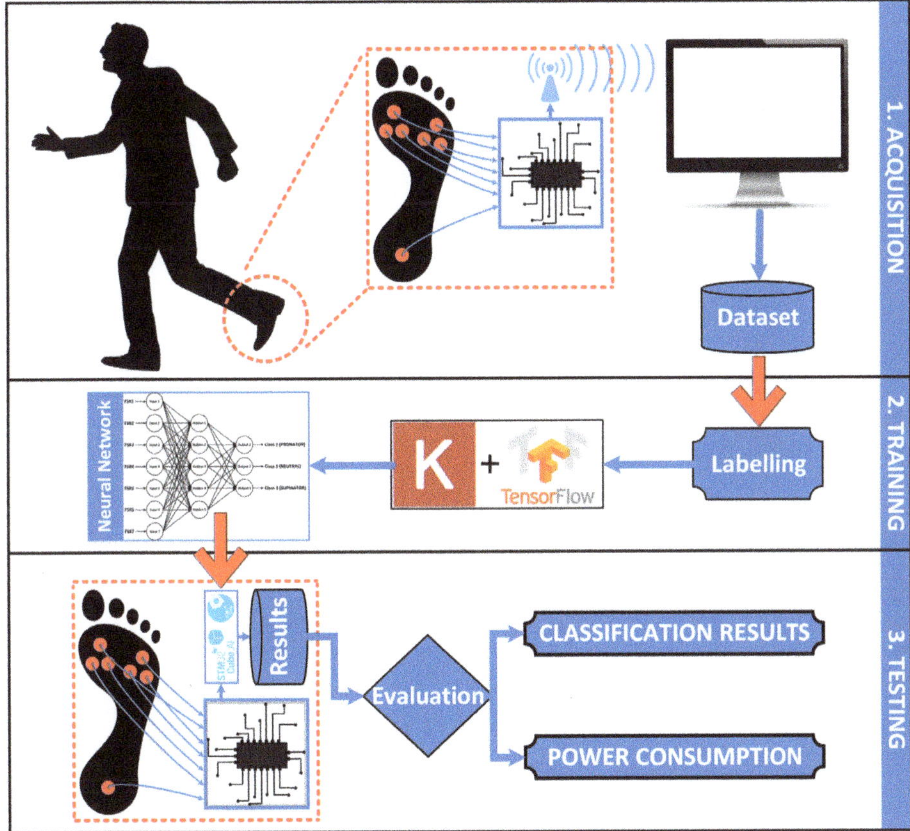

Figure 3. Full system implemented in this work: (**up**) data acquisition phase, (**middle**) training phase and (**bottom**) testing phase.

Once the information is stored, it is important to give further details on the collected database .

2.1.2. Dataset

To obtain a useful database for this work, we need users with different footprint characteristics. To be sure their footprint type has been correctly identified, only previously diagnosed patients have been used. Thus, finally, we recruited six volunteers to acquire the dataset, two users for each type of footprint; that is, two pronators, two supinators and two users with neutral gait. Although these volunteers had been diagnosed previously, we also used a classical pressure platform to verify their footprint type.

Even though the acquisition system (see Figure 3-up) allows the user to configure the sensors acquisition frequency, but the dataset was elaborated with samples taken at 50 Hz (in the next sections we will show that results justify that there is no need for using a higher frequency). With this configuration, the total number of stored footprint samples was approximately 3100, 1020 $\pm$ 90 samples per each type. The results of the data acquisition phase are further detailed in the Results and Discussion section.

2.2. Artificial Neural Network Classifier

Artificial Neural Networks (ANNs) have been used in several works recently to find the relationship between some input data and the desired response at the system output (called the system's inference or classification). This machine learning mechanism is very useful in applications where there are large amounts of input data and the relationship between that data and the expected output cannot be easily appreciated [23]. Thus, after an initial 'training' phase, the ANN is configured with a set of weights at both outputs and inputs of the different neuron layers with which the desired outputs can be obtained. ANNs have been demonstrated to obtain very good results in previous works, especially when used as a supervised machine learning method [24–26].

Their structure, based solely on arithmetic operations (except perhaps in inference phases in classification problems), allows them to be combined with other architectures, making them a fundamental component in several Deep Learning algorithms. It is also possible to create very efficient implementations, which can be optimized for low performance devices, such as low-power microcontrollers [27]. This fact allows acceptable execution times with very low power consumption. In this section we describe the ANN architectures analysed in terms of effectiveness, as well as their performance when embedded in a low-power device.

2.2.1. Architecture Design

An ANN architecture with three layers is used in this gait classification study (see Figure 4). The first layer, that is, the input, contains seven nodes that receive information from the different FSRs, for each footstep. The last layer, the output, consisting of three nodes that return the degree of confidence of a sample to belong to one of the three footprint classes (supinator, pronator or neutral). A final output function called softmax is used. It implements a multi-class sigmoid and is typically used to normalize the results of the network output layer, limiting each output to the 0 to 1 range. The sum of the values of all the output nodes for this function is always 1. This allows us to interpret the output directly as a probability or confidence. Thus, the predicted class would be that with a greater confidence. The intermediate layer, called the hidden layer, is connected to the input and the output layers. Each node of this layer receives the information of all the seven input nodes, processes it and transmits its result to the three nodes in the output layer. Thus it is called a Fully Connected or Dense layer. In this study, architectures with different numbers of nodes in the hidden layer were considered, in order to reduce the complexity of the architecture while maintaining a good classification effectiveness. Therefore we have to look for a high-accuracy network with low complexity to implement it inside a microcontroller as this implies less memory usage and a smaller power consumption.

In this study, TensorFlow https://www.tensorflow.org together with Keras https://keras.io/ have been used to implement, train and test the architectures with the previously detailed dataset (see Figure 3-middle for a global description of the training phase). Tensorflow is a library created by Google for distributed numerical computation, that allows to design, train, evaluate and run models based on neural networks. Keras is a high-level API that simplifies model implementation with Tensorflow, by efficiently managing the connection between model layers and simplifying the Tensorflow code. The resulting models are compatible with STM32Cube utilities, allowing the creation of a C-compiled version that can be embedded in STM32 microcontrollers.

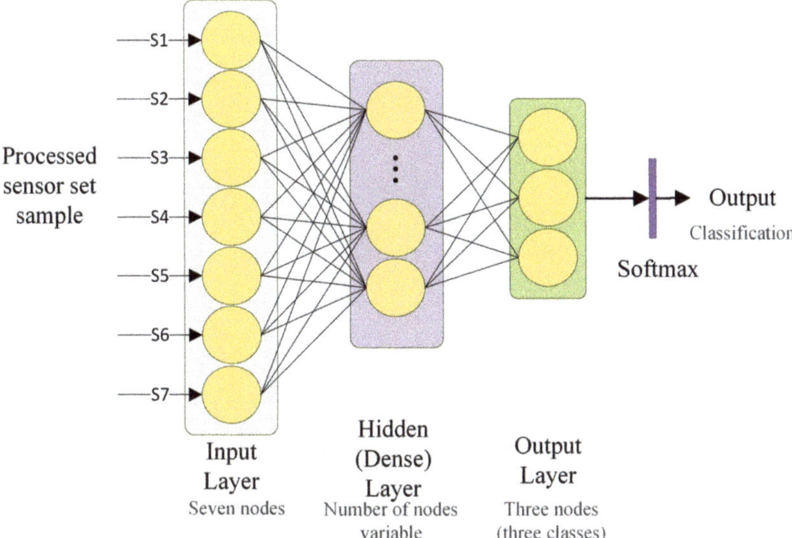

Figure 4. Artificial Neural Network (ANN) architecture used in the study for classification.

2.2.2. Embedded Model Analysis

After the acquisition and training phases (see Figure 3-up and Figure 3-middle, respectively), the ANN is integrated into the embedded system in order to evaluate whether it provides the same effectiveness results as the original implementation on a general-purpose computer, as well as the improvements in energy consumption. For that purpose, STM32CubeIDE development environment https://www.st.com/en/development-tools/stm32cubeide.html was used. It provides tools for power consumption analysis when using different components and modules. Additionally, their STM32Cube.AI expansion plug-in https://www.st.com/en/embedded-software/x-cube-ai.html allows to generate C-compiled versions of pre-trained Neural Networks models, optimized for STM32 microcontrollers.

To verify that the model effectiveness is maintained after conversion to a C-compiled version and integration into the microcontroller, we compared the output confidence results for each class with those obtained with the original Keras model. We used the same MCU that the one used for the acquisition phase for the integration and performance analysis of the trained models.

For the power efficiency analysis, two different scenarios were considered (see Figure 5). In the first one, the embedded system collects data from the sensors at 50Hz and sends it via Bluetooth (every 20 ms); the information is received by the host, which processes and classifies it using an external ANN classifier (this scenario is similar than the one used for the acquisition phase, but now the ANN classifier is implemented in the host). In this case, the system spends almost 4ms for each data transmission (sending 56 bytes at 115,200 bauds), which is 20% of the total time; and, moreover, the transmission process takes more than 43 mA of power consumption (much more than the average power consumption of the system).

In the second scenario, the classification process is done inside the embedded system with the internal ANN implementation. The information read from the sensors is processed and stored internally and, only once per step, the ANN classifies the footstep type and, after that, the classification result is transmitted to the host using Bluetooth. Two main improvements are obtained in this second scenario—first, the transmitted data size is significantly reduced (from 56 bytes to 3 bytes) and, thus, the transmission time is much lower that in the first scenario (0.2 ms); and, second, the transmission is only done after each classification (only one per step), so the number of transmissions is much less than that in the first scenario. However, the number of transmissions depends on the gait cadence of the user, so it must be studied in depth.

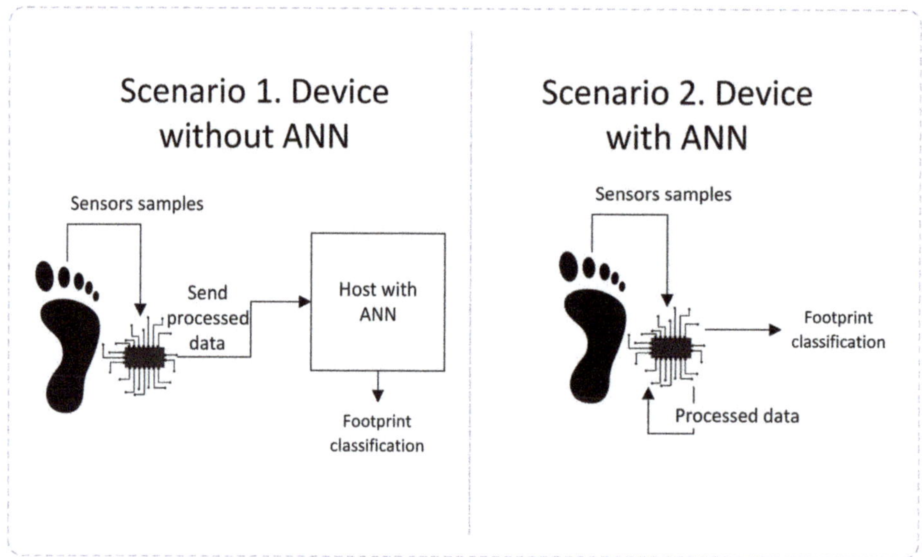

Figure 5. Two scenarios proposed for power consumption analysis.

To easily understand both scenarios, the implemented firmwares are described. The first one (see Figure 6-left) was used to acquire the data for the database and it was also used as "scenario 1" for the final power-consumption comparison. The second one (see Figure 6-right) is used for the embedded ANN implementation and hence it corresponds to "scenario 2" in the testing phase.

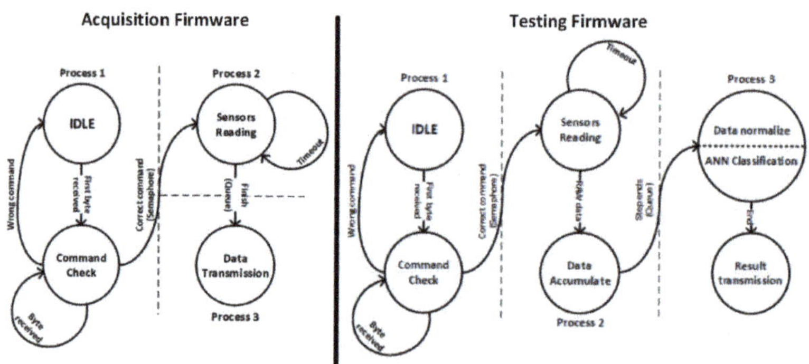

Figure 6. Implemented Firmwares for the acquisition (**left**) and testing phase (**right**).

The firmware implemented in the MCU for the acquisition phase is a FreeRTOS based implementation that allows a bi-directional serial communication with the bluetooth module (process 1 for reception and process 3 for transmission) and a periodic sensor readings with data transformation (process 2), using a binary semaphore and a queue to communicate between these process. So, in the acquisition phase, there is no recording inside the MCU memory; but, after each reading, the data is packed and sent to an external computer where the information is stored in a database.

And, for the testing phase, the implemented firmware is also FreeRTOS based and uses periodic readings, value normalization and performs an ANN classification after each steps ends. The result of this classification is transmitted (once per step) to the external computer.

3. Results and Discussion

In this section, the results obtained at the end of each phase are detailed. For the acquisition phase (see Figure 3-up), the collected dataset is presented. For the training step (see Figure 3-middle), the classification results are detailed and, finally, the embedded model accuracy and the power consumption study are presented as results of the testing phase (see Figure 3-bottom).

3.1. Dataset

The dataset used in this study was split for training and assessment purposes. We used the Hold-out technique, by randomly selecting a sample subset for the training of the models, and using the remaining subset to validate the model performance. A subset with the 85% of dataset samples was used for training, while the remaining 15% subset was used for evaluation. The distribution was made to ensure that there was a balanced percentage of each type of footprint in both subsets. Table 1 shows the distribution.

Table 1. Dataset distribution for each subset.

Subset	Neutral	Pronator	Supinator	Total
Total	1067	1129	928	3124
Training	917	955	784	2656
Test	150	174	144	468

3.2. ANN Model Assessment

This section presents the trained ANN architectures, their implementation and training specifications and the effectiveness evaluation results.

3.2.1. ANN Architectures and Parameters

We analyse the architecture introduced in Section 2.2.1 with different numbers of nodes in its hidden layer. In previous studies, 5 hidden nodes were used in the hidden layer, based on [28]. For the current analysis, we assess the effectiveness by reducing the number of nodes looking for an improvement of classification times and power consumption reduction. Sigmoid function was used as activation function for the nodes in the hidden layer, while Rectified Linear Unit (ReLU) function was used for the output layer nodes. The model training was performed with a learning rate of 0.001, a batch size of 8 and 75 epochs. The used optimizer was a Root Mean Square Prop (RMSProp).

3.2.2. Effectiveness Results

We compared the effectiveness using different metrics—accuracy, sensitivity (also named macro recall), specificity, macro precision and macro F1-score [29]. This last metric measures the relation is the harmonic mean of macro precision and macro recall.

$$Specificity = \sum_c \frac{TN_c}{TN_c + FP_c}, c \in classes \tag{1}$$

$$Precision_m = \sum_c \frac{TP_c}{TP_c + FP_c}, c \in classes \tag{2}$$

$$Recall_m(sensitivity) = \sum_c \frac{TP_c}{TP_c + FN_c}, c \in classes \tag{3}$$

$$F1 - score_m = 2 * \frac{precision_m * recall_m}{precision_m + recall_m}, \tag{4}$$

where m index refers to macro metric and $classes = \{Pronator, Neutral, Supinator\}$. The term TP_c refers to the number of samples with class c that were classified correctly as c by the trained model. TN_c denotes the number of samples with a different class of c that were not classified as c. The term FP_c determines the set of samples with different class of c that were classified as c by the model and, finally, FN_c refers to the set of samples with class c that were wrongly classified as other different class.

The results obtained with each architecture are shown in Table 2. As can be seen, the reduction in the number of nodes not only maintains effectiveness, but also improves when compared to larger models. The greatest effectiveness is achieved with three nodes in the hidden layer. This may be due to the fact that this hidden layer reduction diminishes the so-called over-fitting phenomenon [30], preventing the model from adjusting too closely to the particular characteristics of the used training subset. The architecture, however, can assimilate enough footprint characteristics even with only two hidden nodes with slightly worse effectiveness.

Table 2. Metrics results with different numbers of nodes in the hidden layer. The model trained with only one node in its hidden layer is not able of distinguish one footprint class, so specificity and sensitivity cannot be obtained for this case.

Nodes in Hidden Layer	Accuracy	Precision	F1-Score	Specificity	Sensitivity
Five nodes	0.987	0.987	0.994	0.987	0.987
Four nodes	0.994	0.994	0.997	0.994	0.994
Three nodes	0.996	0.996	0.998	0.996	0.996
Two nodes	0.991	0.992	0.996	0.992	0.992
One node	0.678	0.653	0.842	-	-

3.3. Embedded System Results

In this section, the results obtained from the embedded models running in a low power STM32L476RG board are presented. Three aspects were analysed in relation to the embedded device performance. First, we assessed the accuracy of the C-compiled model obtained with CUBE-AI package extension. Second, we estimated the inference time, that is, the execution times obtained when the embedded model classifies a sample. Finally, we calculated the consumption of the device for the two scenarios established in the Methods section, that is, when the device has the integrated ANN model and when it only sends the data to an external computer with higher computing performance and less power usage limitations.

3.3.1. Embedded Model Accuracy

We analysed the similarity of the outputs from the Keras model and those from its C-compiled version. For this purpose, we assessed the differences on the inference outputs of the models. This was obtained by calculating the relative L2 error:

$$e = \frac{\| F_{generated} - F_{original} \|}{\| F_{generated} \|}, \tag{5}$$

where $F_{generated}$ is the flattened array of the generated model last output layer and $F_{original}$ the flattened array of the original model. In other words, we compare the reliability results returned by the last layer of the two model implementations, prior to classification.

Results for each model are presented in Table 3. We showed the results when each C-compiled model was compressed to occupy less flash memory in the microcontroller. Compression is carried out using a weight sharing-based algorithm. A clustering technique (K-means) is used to calculate values centroids for the layer weights and bias. The compression with factor x4 uses 256 centroids codified on 8 bits, while the compression with factor x8 uses 16 centroids codified on 4bits. In most of the models the L2 error was very low, under 6.8×10^{-7}, which implies the C-compiled models maintain a very close classification behaviour. It should be noted that, for the models with the highest number of nodes in the hidden layer, their more compressed version provides reliability values relatively further from the corresponding model in Keras. Increasing the complexity of the hidden, fully connected layer may have caused this effect. The results obtained may also have been influenced as a consequence of the ov-erfitting effect mentioned above, which could imply an improvement in effectiveness, due to the fact that specific features to classify particular cases of the training set could be forgotten. This may have occurred with the four hidden node compressed model, which has improved its accuracy over the original Keras model. However, the best results continue to be found with the model with three nodes in the hidden layer, obtaining an L2 lower than 1.0×10^{-8}.

Table 3. L2 error for each model (trained model vs. generated c-model) with different compression factors.

Nodes in Hidden Layer	Not-Compressed	x4 Compression	x8 Compression
Five nodes	6.816×10^{-8}	6.816×10^{-8}	9.629×10^{-2}
Four nodes	2.586×10^{-7}	2.586×10^{-7}	2.012×10^{-1}
Three nodes	9.099×10^{-8}	9.099×10^{-8}	9.099×10^{-8}
Two nodes	4.346×10^{-7}	4.346×10^{-7}	4.346×10^{-7}
One node	7.191×10^{-7}	7.191×10^{-7}	7.191×10^{-7}

3.3.2. Execution Times

We estimated the time spent on classifying a sample, that is, the execution time for one ANN classification. This value is important to determine the power consumption for the process 3 in the scenario 2 (see Figure 6-right). We also analysed the results for each compressed model version. The results, which can be seen in Table 4, show that there is a slight variation in power consumption as the number of nodes decreases. The compressed version of each model does not seem to disturb the execution times, except again in the case of the models with the highest number of nodes in the hidden layer, which take longer to perform a classification. Considering the previous results, the architecture with the greatest effectiveness for this problem and with a good classification time is the one with three nodes in its hidden layer, which in turn can be compressed by a factor $\times 8$ without altering performance.

Regarding the power consumption analysis the ANN with three nodes in the hidden layer and a $\times 8$ compression is used.

3.3.3. Power Consumption Analysis

As detailed in previous sections, the main differences of both analysed scenarios are the communications frequency and the amount of data transmitted (see Figure 5).

Table 4. Estimated execution times per classification (milliseconds), of each model integrated and running in STM32L476RG board.

Nodes in Hidden Layer	Not-Compressed	×4 Compression	×8 Compression
Five nodes	0.071	0.071	0.075
Four nodes	0.067	0.067	0.070
Three nodes	0.061	0.061	0.061
Two nodes	0.056	0.056	0.056
One node	0.052	0.052	0.052

In the first scenario (see Figure 5-left), every 20 ms (50 Hz reading frequency) the embedded system takes: 0.07 ms reading the sensors' values (that consumes 6.1 mA), 3.9 ms transmitting via Bluetooth (that consumes 43.16 mA) and the rest of the time (16.03 ms) in sleep mode (that consumes only 18 µA). So, using an average button battery of 125 mAh capacity, the system has a battery life of 14 h.

In the second scenario, the calculation is not that easy because the system only transmits information once per step. So, two possibilities are evaluated: first, sensors are read but the step is not finished; and, second, sensors are read and the step is finished. In the first possibility (step is not end yet): sensors are read and values are accumulated (in this case, the embedded system does not classify and does not transmit). In the second possibility: sensors are read, values are accumulated, the final amount of data for the full step is normalized and classified using the ANN; and, finally, the classification result is transmitted.

Both possibilities are very different in the power-consumption analysis—the second one spends much more power that the first one because of the ANN classification and the transmission.

Moreover, if we compare the power-consumption between the first scenario (always transmitting) and the second possibility of the second scenario (transmission only once per step), there is a big difference too—in the second scenario, only one data transmission per step is done and the time spent in the transmission is less than in the first scenario because the amount of data transmitted is much lower (3 bytes versus 56 bytes), taking only 0.2 ms in the transmission process.

So, evaluating the second scenario, 0.07 ms are spent for sensors' reading (6.1 mA), 0.061 ms are spent for the ANN classification (255.1 µA), 0.2 ms are spent for the transmission (43.16 mA) and the rest of the time (19.66 ms) the system is in sleep mode (18 µA). However, if the step is not ended, the system does not transmit and there is no classification process (only periodic sensors' reading); so, during the time spent in these two phases, the system is in sleep mode too.

The first scenario is relatively easy to evaluate in the power-consumption study, but the second scenario is much more difficult because it depends on the user's gait cadence. So, in order to obtain a more accurate power consumption study for it, the gait cadence of the user must be evaluated. Using the information obtained after the study done in [31], we can observe than a gait cadence less than 100 steps/min corresponds to a low intensity (walking), a cadence between 100 and 130 steps/min corresponds to medium intensity (jogging) and a cadence higher than 130 steps/min corresponds to high intensity (running).

Thus, in our case, we have analysed the power consumption with a sensors' reading frequency fixed at 50 Hz and cadence values between 30 steps/min and 160 steps/min, obtaining the results presented in Table 5. The first column indicates the gait cadence in number of steps per minute and varies from 30 to 160; in the second one, the time spent for each step (in seconds) for each gait cadence is detailed; the third one calculates the total number of samples collected for each step using the data from the previous columns and using only one foot (as one instrumented insole collects information from one foot).

Instead of evaluating the power consumption of both possibilities in the second scenario (step ends or not), for this power-consumption estimation we assume that all the sensors' readings imply a classification step and a data transmission and, depending on the number of samples for each step, the consumption of those processes are multiplied by a factor (between 0 and 1) that indicates the proportion of transmissions depending on the gait cadence. For example, if we always transmit, this factor will be 1; or, if we have 10 samples per step, we transmit only 10% times, so this factor is 0.1. So, the fourth column of Table 5 represents the result of the power consumption of the classification and transmission processes (43.16 mA approx.) multiplied by this factor. Finally, the fifth and sixth columns indicate the average consumption of the full system (in μA) and the final battery life (in hours), respectively.

Table 5. Number of steps, transmissions and power consumption varying the gait cadence of the user.

Cadence (Steps/min)	Sec. Spent for Each Step	# Samples for Each Step (1 foot)	Tx Power Consumption for Each Reading (μA)	System Average Consumption (μA)	Battery Life (h)
30	2.00	50.00	0.86	75.11	1639
40	1.50	37.50	1.15	85.26	1488
50	1.20	30.00	1.44	95.41	1297
60	1.00	25.00	1.73	105.57	1171
70	0.86	21.43	2.01	115.37	1071
80	0.75	18.75	2.30	125.53	984
90	0.67	16.67	2.59	135.68	909
100	0.60	15.00	2.88	145.84	845
110	0.55	13.64	3.17	155.64	792
120	0.50	12.50	3.45	165.80	743
130	0.46	11.54	3.74	175.95	709
140	0.43	10.71	4.03	186.11	670
150	0.40	10.00	4.32	196.26	636
160	0.38	9.38	4.60	206.06	605

Hence, as can be seen in Table 5, in the worst case (highest gait cadence evaluated) the battery life exceeds 25 days. So, it improves the battery life of the first scenario by more than 43 times (an improvement of 4321%). This comparison can be observed in Figure 7.

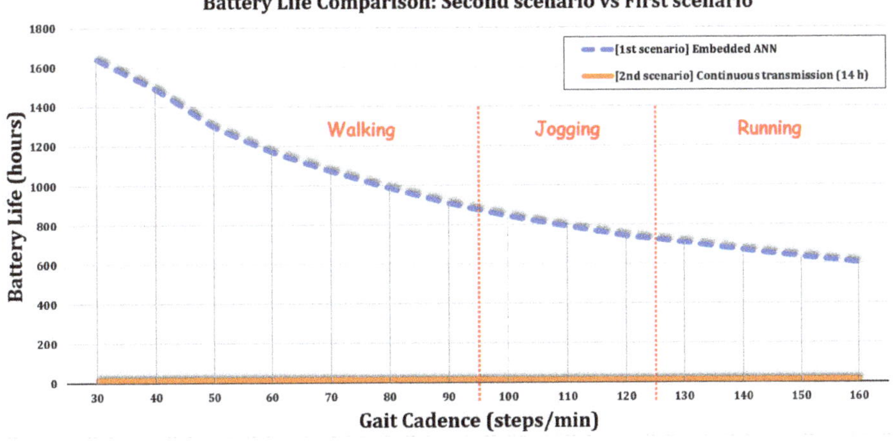

Figure 7. Battery life (in hours) with the power consumption estimation for each scenario.

The results presented in Figure 7 show that the higher the gait cadence, the lower the battery life is. Despite this, the life of the used battery (125 mAh) even for cases in which the user is running high enough to allow a biomechanical gait study without recharges.

4. Conclusions

In this work, a performance analysis of a low-power footwear insole for the detection of abnormal foot postures is presented. The device implements an embedded Machine Learning model based on ANN for real-time footprint type inference. The inputs of the model consist of average FSR measures obtained during a footstep, and the outputs correspond to the three gait types described in the Introduction section—pronator, supinator and neutral.

First, a model study was performed. The effectiveness of the ANN architecture, consisting of three neural layers, was assessed using a different number of nodes in the hidden layer. The architecture with three nodes obtained the best results, with effectiveness metrics above 99.6%. The architectures with a greater number of nodes showed slightly less classification ability, possibly due to overfitting the training dataset.

Finally, as the main point of the study, a complete analysis of the classifier performance has been performed when it is integrated into a low-power embedded device. The L2 error obtained when comparing the Keras and the C-compiled model outputs showed that the conversion does not have a significant impact on the effectiveness of the model, even when the model is compressed, thus saving memory space on the microcontroller. This can be an important aspect if we intend to include other models with different functionalities in the device in future works. Regarding the inference execution times, the best model is able to classify a footstep sample in 0.61 ms, even when it is compressed. This is much less than the time needed to read a sample of the insole sensors, thereby achieving real-time execution. Based on this, and considering that the classifier execution and result transmission only take place when a full step is performed, the battery life estimation is over 25 days (considering the higher gait cadence).

Author Contributions: Conceptualization, M.D.-M., D.G.-G. and A.C.-B.; methodology, F.L.-P., M.D.-M. and A.C.-B.; software, F.L.-P. and D.G.-G.; validation, F.L.-P., M.D.-M. and A.C.-B.; investigation, F.L.-P., D.G.-G. and M.D.-M.; resources, F.L.-P. and M.D.-M.; writing–original draft preparation, F.L.-P., D.G.-G. and M.D.-M.; writing–review and editing, D.G.-G. and A.C.-B.; supervision, M.D.-M. and A.C.-B.; funding acquisition, A.C.-B. All authors have read and agreed to the published version of the manuscript.

Funding: This research was funded by the Robotics and Computer Technology Lab. (RTC) of the Universidad de Sevilla, Spain.

Acknowledgments: This work has been supported by the Robotics and Computer Technology Lab. (RTC) of the Universidad de Sevilla, Spain.

Conflicts of Interest: The authors declare no conflict of interest.

Abbreviations

The following abbreviations are used in this manuscript:

OS	Operating System
ANN	Artificial Neural Network
FSR	Force Sensitive Resistors
BLE	Bluetooth Low Energy
MCU	Microcontroller
ReLU	Rectified Linear Unit

J. Low Power Electron. Appl. **2020**, *10*, 14

References

1. Thomas, M.J.; Roddy, E.; Zhang, W.; Menz, H.B.; Hannan, M.T.; Peat, G.M. The population prevalence of foot and ankle pain in middle and old age: A systematic review. *Pain* **2011**, *152*, 2870–2880. [CrossRef]
2. Menz, H.B. Chronic foot pain in older people. *Maturitas* **2016**, *91*, 110–114. [CrossRef] [PubMed]
3. Menz, H.B.; Dufour, A.B.; Riskowski, J.L.; Hillstrom, H.J.; Hannan, M.T. Association of planus foot posture and pronated foot function with foot pain: The Framingham foot study. *Arthritis Care Res.* **2013**, *65*, 1991–1999. [CrossRef]
4. Buldt, A.K.; Allan, J.J.; Landorf, K.B.; Menz, H.B. The relationship between foot posture and plantar pressure during walking in adults: A systematic review. *Gait Posture* **2018**, *62*, 56–67. [CrossRef] [PubMed]
5. Perry, J.; Davids, J.R. Gait analysis: Normal and pathological function. *J. Pediatr. Orthop.* **1992**, *12*, 815. [CrossRef]
6. Buldt, A.K.; Forghany, S.; Landorf, K.B.; Levinger, P.; Murley, G.S.; Menz, H.B. Foot posture is associated with plantar pressure during gait: A comparison of normal, planus and cavus feet. *Gait Posture* **2018**, *62*, 235–240. [CrossRef]
7. Razeghi, M.; Batt, M.E. Foot type classification: A critical review of current methods. *Gait Posture* **2002**, *15*, 282–291. [CrossRef]
8. Frelih, N.G.; Podlesek, A.; Babič, J.; Geršak, G. Evaluation of psychological effects on human postural stability. *Measurement* **2017**, *98*, 186–191. [CrossRef]
9. Morris, S.J.; Paradiso, J.A. Shoe-integrated sensor system for wireless gait analysis and real-time feedback. In Proceedings of the Second Joint 24th Annual Conference and the Annual Fall Meeting of the Biomedical Engineering Society, Houston, TX, USA, 23–26 October 2002; Volume 3, pp. 2468–2469.
10. Bamberg, S.J.M.; Benbasat, A.Y.; Scarborough, D.M.; Krebs, D.E.; Paradiso, J.A. Gait analysis using a shoe-integrated wireless sensor system. *IEEE Trans. Inf. Technol. Biomed.* **2008**, *12*, 413–423. [CrossRef]
11. Shu, L.; Hua, T.; Wang, Y.; Li, Q.; Feng, D.D.; Tao, X. In-shoe plantar pressure measurement and analysis system based on fabric pressure sensing array. *IEEE Trans. Inf. Technol. Biomed.* **2010**, *14*, 767–775.
12. Wahab, Y.; Mazalan, M.; Bakar, N.; Anuar, A.; Zainol, M.; Hamzah, F. Low power shoe integrated intelligent wireless gait measurement system. *J. Phys. Conf. Ser. IOP Publ.* **2014**, *495*, 13–14. [CrossRef]
13. Crea, S.; Donati, M.; De Rossi, S.M.M.; Oddo, C.M.; Vitiello, N. A wireless flexible sensorized insole for gait analysis. *Sensors* **2014**, *14*, 1073–1093. [CrossRef] [PubMed]
14. Talib, N.; Rahman, M.; Najib, A.; Noor, M. Implementation of Piezoelectric Sensor in Gait Measurement System. In Proceedings of the 2018 8th IEEE International Conference on Control System, Computing and Engineering (ICCSCE), Penang, Malaysia, 23–25 November 2018; pp. 20–25.
15. Oshimoto, T.; Abe, I.; Kikuchi, T.; Chijiwa, N.; Yabuta, T.; Tanaka, K.; Asaumi, Y. Gait Measurement for Walking Support Shoes with Elastomer-Embedded Flexible Joint. In Proceedings of the 2019 IEEE/SICE International Symposium on System Integration (SII), Paris, France, 14–16 January 2019; pp. 537–542.
16. Ngamsuriyaroj, S.; Chira-Adisai, W.; Somnuk, S.; Leksunthorn, C.; Saiphim, K. Walking gait measurement and analysis via knee angle movement and foot plantar pressures. In Proceedings of the 2018 15th International Joint Conference on Computer Science and Software Engineering (JCSSE), Nakhonpathom, Thailand, 11–13 July 2018; pp. 1–6.
17. Dominguez-Morales, J.P.; Rios-Navarro, A.; Dominguez-Morales, M.; Tapiador-Morales, R.; Gutierrez-Galan, D.; Cascado-Caballero, D.; Jimenez-Fernandez, A.; Linares-Barranco, A. Wireless sensor network for wildlife tracking and behavior classification of animals in Doñana. *IEEE Commun. Lett.* **2016**, *20*, 2534–2537. [CrossRef]
18. Henkel, J.; Pagani, S.; Amrouch, H.; Bauer, L.; Samie, F. Ultra-low power and dependability for IoT devices (Invited paper for IoT technologies). In Proceedings of the Design, Automation & Test in Europe Conference & Exhibition (DATE), Lausanne, Switzerland, 27–31 March 2017; pp. 954–959.
19. Andres-Maldonado, P.; Ameigeiras, P.; Prados-Garzon, J.; Ramos-Munoz, J.J.; Lopez-Soler, J.M. Optimized LTE data transmission procedures for IoT: Device side energy consumption analysis. In Proceedings of the 2017 IEEE International Conference on Communications Workshops (ICC Workshops), Paris, France, 21–25 May 2017; pp. 540–545.

20. Deepu, C.J.; Heng, C.H.; Lian, Y. A hybrid data compression scheme for power reduction in wireless sensors for IoT. *IEEE Trans. Biomed. Circuits Syst.* **2016**, *11*, 245–254. [CrossRef]

21. Domínguez-Morales, M.J.; Luna-Perejón, F.; Miró-Amarante, L.; Hernández-Velázquez, M.; Sevillano-Ramos, J.L. Smart Footwear Insole for Recognition of Foot Pronation and Supination Using Neural Networks. *Appl. Sci.* **2019**, *9*, 3970. [CrossRef]

22. Pineda-Gutiérrez, J.; Miró-Amarante, L.; Hernández-Velázquez, M.; Sivianes-Castillo, F.; Domínguez-Morales, M. Designing a Wearable Device for Step Analyzing. In Proceedings of the 2019 IEEE 32nd International Symposium on Computer-Based Medical Systems (CBMS), Cordoba, Spain, 5–7 June 2019; pp. 259–262.

23. Anderson, D.; McNeill, G. Artificial neural networks technology. *Kaman Sci. Corp.* **1992**, *258*, 1–83.

24. Luna-Perejón, F.; Domínguez-Morales, M.J.; Civit-Balcells, A. Wearable Fall Detector Using Recurrent Neural Networks. *Sensors* **2019**, *19*, 4885. [CrossRef]

25. Nasser, I.M.; Abu-Naser, S.S. Lung Cancer Detection Using Artificial Neural Network. *Int. J. Eng. Inf. Syst.* **2019**, *3*, 17–23.

26. Sadek, R.M.; Mohammed, S.A.; Abunbehan, A.R.K.; Ghattas, A.K.H.A.; Badawi, M.R.; Mortaja, M.N.; Abu-Nasser, B.S.; Abu-Naser, S.S. Parkinson's Disease Prediction Using Artificial Neural Network. *Int. J. Acad. Health Med. Res.* **2019**, *3*, 1–8.

27. Gutierrez-Galan, D.; Dominguez-Morales, J.P.; Cerezuela-Escudero, E.; Rios-Navarro, A.; Tapiador-Morales, R.; Rivas-Perez, M.; Dominguez-Morales, M.; Jimenez-Fernandez, A.; Linares-Barranco, A. Embedded neural network for real-time animal behavior classification. *Neurocomputing* **2018**, *272*, 17–26. [CrossRef]

28. Sheela, K.G.; Deepa, S.N. Review on methods to fix number of hidden neurons in neural networks. *Math. Probl. Eng.* **2013**, *2013*, 425740. [CrossRef]

29. Sokolova, M.; Lapalme, G. A systematic analysis of performance measures for classification tasks. *Inf. Process. Manag.* **2009**, *45*, 427–437. [CrossRef]

30. Finnoff, W.; Hergert, F.; Zimmermann, H.G. Improving model selection by nonconvergent methods. *Neural Netw.* **1993**, *6*, 771–783. [CrossRef]

31. Tudor-Locke, C.; Aguiar, E.J.; Han, H.; Ducharme, S.W.; Schuna, J.M.; Barreira, T.V.; Moore, C.C.; Busa, M.A.; Lim, J.; Sirard, J.R.; et al. Walking cadence (steps/min) and intensity in 21–40 year olds: CADENCE-adults. *Int. J. Behav. Nutr. Phys. Act.* **2019**, *16*, 1–11. [CrossRef]

MDPI
St. Alban-Anlage 66
4052 Basel
Switzerland
Tel. +41 61 683 77 34
Fax +41 61 302 89 18
www.mdpi.com

Journal of Low Power Electronics and Applications Editorial Office
E-mail: jlpea@mdpi.com
www.mdpi.com/journal/jlpea